Field Guide
to
Silkmoths of Illinois

John K. Bouseman
James G. Sternburg

Also by the same authors

Field Guide to Butterflies of Illinois:
Illinois Natural History Survey Manual 9

*Field Guide
to*
Silkmoths of Illinois

John K. Bouseman
James G. Sternburg

Illinois Natural History Survey • Champaign
September 2002
Manual 10

Illinois Natural History Survey, David L. Thomas, Chief
A Division of the Illinois Department of Natural Resources,
Brent Manning, Director

Illinois Natural History Survey
Natural Resources Building
607 East Peabody Drive
Champaign, Illinois 61820

Printed by the authority of the State of Illinois

Photo credits: All photographs are by James G. Sternburg unless otherwise indicated in the figure captions. Copyright of each photo resides with the photographer.

Editors: Thomas E. Rice and Charles Warwick

Dust jacket: Thomas E. Rice

ISBN: 1-882932-06-4

Library of Congress Control Number: 2002108786

Citation:
Bouseman, J.K., and J.G. Sternburg. 2002. Field guide to silkmoths of Illinois. Illinois Natural History Survey Manual 10. x + 97 pp.

Printed with soy ink on recycled and recyclable paper.

Equal opportunity to participate in programs of the Illinois Department of Natural Resources (IDNR) and those funded by the U.S. Fish and Wildlife Service and other agencies is available to all individuals regardless of race, sex, national origin, disability, age, religion, or other non-merit factors. If you believe you have been discriminated against, contact the funding source's civil rights office and/or the Equal Employment Opportunity Officer—IDNR, One Natural Resources Way, Springfield, IL 62702-1271; 217/785-0067; TTY 217/782-9175.

This work is dedicated to those amateur and professional entomologists, past and present, whose endeavors have helped to make this field guide possible.

Illinois County Locations

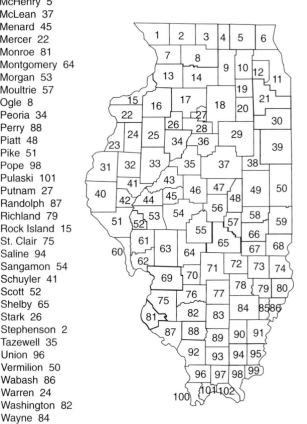

Adams 40
Alexander 100
Bond 70
Boone 4
Brown 42
Bureau 17
Calhoun 60
Carroll 7
Cass 44
Champaign 49
Christian 55
Clark 63
Clay 78
Clinton 76
Coles 66
Cook 11
Crawford 74
Cumberland 67
De Kalb 9
De Witt 47
Douglas 58
Du Page 12
Edgar 50
Edwards 85
Effingham 72
Fayette 71
Ford 38
Franklin 89
Fulton 33
Gallatin 95
Greene 61
Grundy 20
Hamilton 90
Hancock 31
Hardin 99
Henderson 23
Henry 16
Iroquois 39
Jackson 92
Jasper 73
Jefferson 83
Jersey 62
Jo Daviess 1
Johnson 97
Kane 10
Kankakee 30
Kendall 19
Knox 25
Lake 6
La Salle 18
Lawrence 80
Lee 14
Livingston 29
Logan 46
Macon 56
Macoupin 63
Madison 69
Marion 77
Marshall 28
Mason 43
Massac 102
McDonough 32
McHenry 5
McLean 37
Menard 45
Mercer 22
Monroe 81
Montgomery 64
Morgan 53
Moultrie 57
Ogle 8
Peoria 34
Perry 88
Piatt 48
Pike 51
Pope 98
Pulaski 101
Putnam 27
Randolph 87
Richland 79
Rock Island 15
St. Clair 75
Saline 94
Sangamon 54
Schuyler 41
Scott 52
Shelby 65
Stark 26
Stephenson 2
Tazewell 35
Union 96
Vermilion 50
Wabash 86
Warren 24
Washington 82
Wayne 84
White 91
Whiteside 13
Will 21
Williamson 93
Winnebago 3
Woodford 36

Contents

Illinois County Locations (map) vi

Foreword ix

Acknowledgments x

Introduction 1

How to Use This Book 2

Maps of Distribution 2

Economic Considerations 2

Classification 3

Larval Morphology 4

Wing Expansion 8

Mimicry 10

Rearing Saturniids 11

Photographing Insects 12

Saturniid Relatives 14

Species Accounts 15

    Regal Moth or Royal Walnut Moth—Subfamily Ceratocampinae 17

    Pine-devil Moth—Subfamily Ceratocampinae 22

    Imperial Moth—Subfamily Ceratocampinae 24

    Spiny Oakworm Moth —Subfamily Ceratocampinae 28

    Pink-striped Oakworm Moth—Subfamily Ceratocampinae 30

Orange-striped Oakworm Moth—Subfamily Ceratocampinae 32

Rosy Maple Moth—Subfamily Ceratocampinae 34

Bisected Honey Locust Moth—Subfamily Ceratocampinae 36

Honey Locust Moth—Subfamily Ceratocampinae 38

Buck Moth—Subfamily Hemileucinae 41

Nevada Buck Moth—Subfamily Hemileucinae 43

Io Moth—Subfamily Hemileucinae 45

Polyphemus Moth—Subfamily Saturniinae 48

Luna Moth—Subfamily Saturniinae 53

Promethea Moth or Spicebush Silkmoth—Subfamily Saturniinae 58

Tulip Tree Silkmoth—Subfamily Saturniinae 64

Cecropia Moth or Robin Moth—Subfamily Saturniinae 69

Columbia Silkmoth—Subfamily Saturniinae 75

Ailanthus Silkmoth—Subfamily: Saturniinae 80

Glossary 83

Species Checklist 85

Additional Reading 87

Index 93

Foreword

The authors have been encouraged and heartened by the reception of our *Field Guide to Butterflies of Illinois*. Although very authoritative and comprehensive treatments existed for the butterflies of North America, we have found that an audience existed for regional manuals. Faunal works of limited and perhaps arbitrary scope reduce for the casual observer the "noise" generated by extraneous faunal elements in groups of confusingly similar species that can be difficult to determine.

In this book we treat the often spectacularly large and beautiful moths known as the imperial moths or silkmoths. Both the adults and larvae of these insects have long attracted the attention of naturalists, scientists, artists, schoolchildren, and people in general.

We have recently become aware of reports that these heretofore rather common insects are suffering catastrophic population declines. These losses in some parts of the country have been atttributed to the depredations of a parasitic fly, *Compsilura concinnata*, which was imported and introduced into North America from Europe as a component of biological control programs targeted at the gypsy moth and other forest pest species. We are saddened by this calamitous development and further dismayed by the prospect of the possible extermination of these awesomely beautiful creatures by an introduced generalist predator.

John K. Bouseman and James G. Sternburg

Acknowledgments

We are indebted to many persons for their interest in the production of this book and for the kind aid they provided to us in the course of its preparation. Tim Cashatt and Jim Wiker helped us search the collection of the Illinois State Museum for records of moths. Jim Wiker also helped by allowing us to photograph specimens from his magnificent personal collection of Lepidoptera. We enjoyed the hospitality of the Richardson Wildlife Foundation where we were able to work on the manuscript away from the usual daily distractions of life in Urbana. For this we thank Edward J. Richardson, president of the foundation, and Terry Moyer, vice president and resident manager of the facility. Carie Nixon prepared excellent line drawings for the book. University of Illinois entomologists Jim Appleby and Phil Nixon made some much-needed slides available to us, as did James Tuttle of Tuscon, AZ. The manuscript was carefully typed by Dottie Nadarski. Liane Suloway prepared the distributional maps. As always, we were enormously benefited by the superb redactional skills of our editors at the Natural History Survey, Charles Warwick and Thomas Rice. INHS Head Librarian Elizabeth Wohlgemuth and staff librarians JoAnn Jacoby and Jessica Beverly cheerfully sought out elusive references and rare volumes. May Berenbaum and Gilbert Waldbauer of the University of Illinois Department of Entomology and Tim Cashatt of the Illinois State Museum provided thorough, helpful reviews of the initial draft of this publication. We heartily thank all of the above for their indispensable aid.

The authors wish to thank their respective parents, Thomas and Catherine Bouseman and Paul and Eva Sternburg, for their encouragement of our early interest in collecting and rearing moths. JKB thanks Tammie and William Bouseman for their aid in the field.

Introduction

The purpose of this field guide is to enable the user to identify to species any member of the moth family Saturniidae found in the state of Illinois. Secondarily, a selection of literature is offered that will serve to introduce the reader to the voluminous published work that exists about these often spectacularly beautiful insects.

Of the 23 native and one introduced species of saturniid moths that are known from North America east of the Mississippi River, all save 6 have been recorded from Illinois. Those missing are species of limited range in either the southern or northeastern United States. Thus, this field guide should serve well throughout most of the eastern United States.

Because saturniid moths are for the most part nocturnal insects of forests and of somewhat obscure habits, they are not observed as frequently as their diurnal relatives such as butterflies and skippers. As a consequence, their distributions tend not to be well known. Indeed, they are generally noticed only when attracted to lights.

We hope that this summary of what is known about these moths in Illinois will challenge readers to add to our somewhat scanty knowledge of them.

How to Use This Book

The user should become familiar with the characteristics that enable one to recognize a saturniid moth. This is especially important in dealing with the smaller species. Most of the larger species are recognizable at a glance because of their size and unique patterns. Comparison with the illustrations of specimens in the text should lead to the rapid identification of any saturniid moth found in Illinois.

Under the individual species accounts, there are descriptions of the adults and the immature stages (larvae and pupae). If there is the possibility of confusion among various species, this is discussed under "Similar Species." There is a discussion of the "Habitats" where the species is likely to be encountered and the "Life Histories" are treated for all species. Our opinion of the geographic distribution is given under "Status." "Remarks" will add various sorts of information that do not fit comfortably under the other rubrics.

Maps of Distribution

The distributions of the silkmoths and royal moths known to occur in Illinois are indicated on the maps that accompany the individual species accounts and are presented in the form of county records. These records have been compiled from specimens in the collection of the Illinois Natural History Survey supplemented by records from the Illinois State Museum, from the personal experience of the authors, and from published records. The main sources of published records are Paul A. Opler's *Distribution of Silkmoths (Saturniidae) and Hawkmoths (Sphingidae) of Eastern North America* and the "Season Summaries" in the newsletters of the Lepidopterists' Society.

The absence of a record from any particular county should not necessarily be interpreted to mean that a species does not occur in that county. It could simply mean that no amateur or professional entomologist has searched for or collected that species in the county. The maps are indicative, not definitive, of the ranges of the species.

Economic Considerations

A few of the species treated in this book have minor status as economic pests through their occasional defoliation of forest trees. Included among such are the oakworms of the genus *Anisota* and the Rosy Maple Moth. These infestations are generally of short duration.

The larvae of the buck moths and the Io Moth are of medical importance because of their vestiture of urticating (stinging) setae. In contact with bare skin, they cause a severe nettling sensation. They should be handled with considerable caution.

Classification

The family Saturniidae, along with eight other families of moths, constitute the superfamily Bombycoidea of the order Lepidoptera. Only two of these nine families, the Saturniidae and the Sphingidae, are well represented in North America. Five species of the subfamily Apatelodinae of the family Bombycidae occur in the eastern United States. All of the other bombycid subfamilies and the other six bombycoid families are of exotic occurrence. They are mainly Palaearctic species of Eurasian distribution.

Family Saturniidae. Saturniids are medium to large moths, most with stout bodies that are densely clothed with fine, hairlike setae. The head is small, closely attached to the thorax. The labial palps are present but small. The proboscis is reduced or absent and not functional; these moths never feed or drink as adults. Compound eyes are often large in nocturnal species, smaller in some of the diurnal species. There are no ocelli. The antennae of most males are quadripectinate, with long rami (branches). An exception occurs in the buck moths of the genus *Hemileuca* where the antennae are bipectinate. Females have either quadripectinate antennae with reduced rami or bipectinate or simple antennae, depending on the

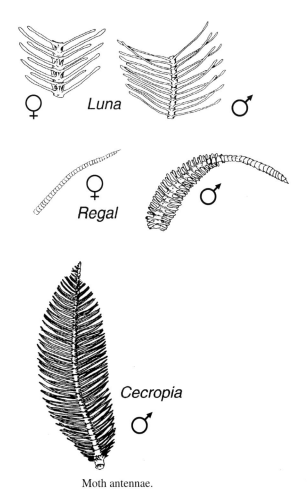

Moth antennae.

species. In all species the forewing has wing vein Cu appearing to be three-branched. The hindwing has Rs widely separated from wing veins Sc + R_1. All saturniids lack a frenulum. To keep the fore and hindwings together in flight, the humeral angle of the hindwing is expanded, and extends forward below the forewing. This is called amplexiform coupling, a condition evolved independently in the butterflies and a few other families of moths. Both nocturnal and diurnal species occur. In Illinois, the male Promethea Moth is diurnal, whereas the female is nocturnal. Both sexes of buck moths of Illinois are diurnal. Two of our *Anisota* species have diurnal males.

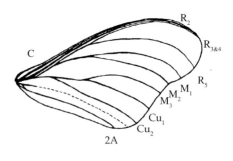

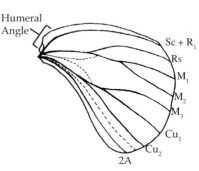

Abbreviation of wing vein terms

A	anal vein	R_1	first branch of radius
C	costa	R_2 to R_5	divisions of the radial sector
Sc	subcosta	M	media
R	radius	M_1 to M_3	divisions of the median vein
Rs	radial sector	Cu	cubitus

Larval Morphology

Saturniid larvae have the typical structures found in the order Lepidoptera. The body is divided into three regions: head, thorax, and abdomen. The head consists of two preoral regions and four postoral segments fused together to form a cranium with stemmata (larval eyes), antennae, labrum, mandibles, maxillae, and labium. The two antennae are very small. On each side of the head there is a semicircular cluster of stemmata, the larval eyes. The paired mandibles, paired maxillae, and the labium—a fusion of second maxillae—are typical of insects with chewing mouthparts. Mandibles are not present in the pupal or adult stages of Lepidoptera (Note: A few primitive moths retain mandibles as adults). The thorax consists of three segments—a prothorax, a mesothorax, and a metathorax—each with a pair of segmented legs. The abdomen has 10 segments, with paired fleshy prolegs on abdominal segments 3, 4, 5, 6, and 10. The prolegs bear minute hooks called crochets that enable the larva to cling to the substrate. The head capsule is hard, whereas the thorax and abdomen are soft and flexible. Spiracles through which air enters and leaves the internal respiratory tracheal system are present on the lateral sides of the prothorax and the first eight abdominal segments.

To aid the reader in identifying the saturniid larvae found in the environment, it is necessary to define and illustrate the cuticular structures and their location on the larvae. Saturniid larvae are easily recognized by their rows of scoli, tuberous outgrowths of the thorax and abdomen. There are none on the head. The few other caterpillars with scoli that might be confused with the

Reared Hickory Horned Devil larva, Pope County.

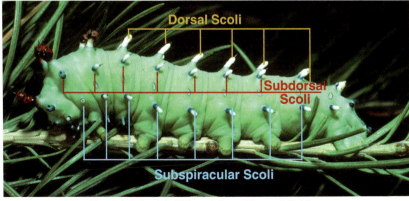

Reared Columbia Silkmoth larva, Lower Michigan.

saturniids are found in some of the butterflies (family Nymphalidae) and several families of moths, including the slug caterpillar moths (family Limacodidae), and some of the measuring worms (family Geometridae). Slug caterpillars have spiny scoli extending outward along the lateral margins of the segments. Some nymphalid larvae have numerous scoli, including some on the head; their arrangement differs from that of the saturniids. Geometrid larvae have a reduced number of ventral prolegs, unlike the saturniids, so they are easily distinguished from saturniids.

Scoli (singular scolus) are sclerotized spiny outgrowths of the integument. The form can be a tubercle with a seta or with a spine, often branched. Scoli of the subfamily Hemileucinae are wartlike with piercing bristles that release and inject an irritating and allergenic substance when the larva is attacked by a predator. Scoli on the attacine Saturniinae are knoblike, or rounded, and often colorful, with many spines; they lend a disruptive appearance to the larva. Scoli are sometimes hornlike, more or less elongate, threatening in appearance but actually harmless. This type occurs on many of the royal moths (subfamily Ceratocampinae.) Scoli of the first instar larvae are usually larger in proportion to body size than in the later instars, in some instances strikingly so. In some larvae, for example in larvae of *Actias luna* and *Antheraea polyphemus*, the scoli of the mature larvae are very small and easily overlooked.

Scoli in the Saturniidae are arranged in longitudinal rows of one per thoracic and abdominal segment. Their locations are dorsal, subdorsal, and subspiracular (or lateral). Thus, there are paired dorsal scoli, and on each side a subdorsal and a subspiracular scolus. On the thoracic segments there may be a lateroventral scolus.

In addition to the scoli, there are in some species small raised cuticular granules, scattered more or less densely over the body surface, as in species of the genus *Anisota*. All larvae have setae, with definite locations, some occurring alone in the first instar and called primary setae. Others appear in the later instars. In most saturniids all setae are short and not at once noticeable. However, there are exceptions; for example, in larvae of *Eacles imperialis*, where the setae are very long and somewhat dense, giving the insect a hairy look.

Subfamily Ceratocampinae. Members of the subfamily Ceratocampinae are known as the royal moths. The antennae are quadripectinate in the basal one-half to two-thirds only, the outer part simple. The abdomen is as long as the hindwings or longer. At rest, most of our species hold the wings in a flexed position over the abdomen. An exception is *Eacles*, which rests with the wings widespread. Larvae have at least one pair of thoracic scoli modified as long hornlike structures. Pupation is in the soil, with no cocoon. Pupae are smooth or rugose, or spiny with a prominent cremaster.

Subfamily Hemileucinae. Here are included the buck moths and Io Moth. The

Larval Morphology

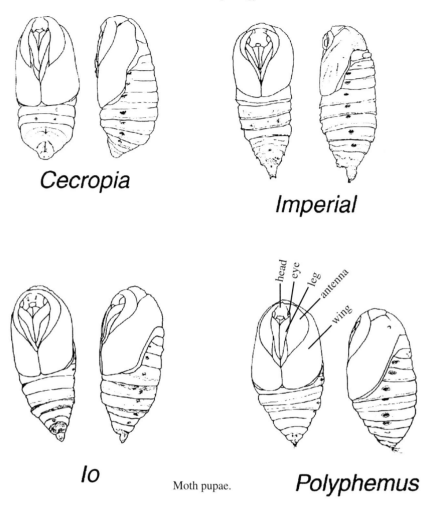

Moth pupae.

antennae of males are either bipectinate or quadripectinate all the way to the tip. If quadripectinate, the outer pair of rami touch the inner pair of the next segment. Females have either simple or bipectinate antennae with short rami. Larvae have scoli with urticating spines or hairlike setae, releasing a toxin when broken that is extremely irritating, causing a burning sensation and rash. Both sexes of the two buck moths are diurnal, while those of the Io Moth are nocturnal. Larvae of buck moths and the Io Moth spin cocoons in or under the leaf litter below the larval host plant.

Subfamily Saturniinae. These are large to very large moths, often very colorful, or cryptic, or with prominent eyespots on the wings. The antennae of both sexes are quadripectinate to the tips, with long rami in the males, and much shorter rami in the females. At rest, depending on the species, the wings are held open but not flexed (Luna Moth), or simply folded together and not flexed above the body, resembling the true butterflies in this habit. The adults of most are nocturnal. An exception among the Illinois species is found in the Promethea Moth, where the males are diurnal in flight and mating, the females diurnal in mating, and nocturnal in flight and oviposition. Larvae have well-developed scoli, both lateral and dorsal in position. Our species do not have urticating hairs, although these are present on a few species found elsewhere. Larvae of the Cecropia Moth may attain a body weight of 20 grams or more. Larvae spin strong silken cocoons, sometimes on the larval host, but often on a nearby support where the cocoon can be hidden from predators. The structure of the cocoon varies, some being relatively simple with no escape valve, to others that are complex and double-walled with an escape valve.

Wing Expansion

Upon escape from the cocoon, the teneral (recently molted, still soft) adult climbs quickly up a branch or tree trunk to find a spot that will allow the rapidly expanding, thick, small, soft wings to hang down so that they can properly assume the adult wing shape. Wing expansion is due to the positive pressure of blood pumped into the wings, forcing the still-soft tissues to stretch. Within a short time (15 minutes, more or less), the wings assume their final size. In the case of the Luna Moth, the long tails are the last part to fully expand. Biochemical action then hardens the wings. We show a series of photographs (pages 8–9) of the wing expansion of a Luna Moth, from the time of emergence from the cocoon to the final positioning of the fully formed and sclerotized (hardened) wings. This phase of

1

2

3

Wing Expansion

4

5

6

7

8

9

the moth's life is in many ways the most critical. If the wings do not expand properly, the insect will be unable to fly normally. If the moth is found by a predator at this stage while still soft, it is defenseless, easy prey for a predator.

Mimicry

The diurnal males of *Callosamia promethea* are largely black, and in flight resemble any of our blackish swallowtails, including the toxic Pipevine Swallowtail (*Battus philenor*). The resemblance includes, to our eye, their wingbeat frequency, their speed, and their general appearance in flight. A male Promethea Moth in search of a pheromone-releasing female has a deliberate flight, not stopping in search of nectar or other energy sources, and, from our experience, can be mistaken for a swallowtail. The behavioral characteristics are an integral part of the resemblance, not at once apparent when viewing a dead museum specimen. Nevertheless, field experiments with live Promethea Moths painted to resemble palatable and unpalatable insects have convincingly shown the survival value of appearing to be an insect known to be toxic and unpalatable to birds. Captive naïve birds have been found to eat Promethea Moths without ill effects, thus the mimicry must be Batesian (see Glossary for definition).

It has been suggested (see Ferguson 1971) that the males of *Anisota virginiensis* and *A. senatoria* may be wasp or bee mimics. The day-flying males, when gathered in numbers over caged virgin females of their species, gave the impression of swarming bees. Whether or not this is mimicry has not been determined.

The possibility of Müllerian (see Glossary) mimicry by *Hemileuca maia maia* and *H. nevadensis* has been suggested (Ferguson 1971; Tuskes et al. 1996) because of their similarity where their ranges overlap.

An example that may be allopatric Müllerian mimicry occurs between *Dryocampa rubicunda* and the Neotropical Royal Moth *Psilopygida apollkinairei*, not closely related, but very similar in appearance. Within the range of *D. rubicunda*, the notodontid moth *Hyparpax aurora* is nearly identical. The noctuid *Schinia florida* is similar, but with the colors reversed.

Pink Prominent Moth male (*Hyparpax aurora*), Family Nolotontidae, Cook County.

Primrose Moth (*Schinia florida*), Family Noctuidae, Lake County, Indiana.

Primrose Moth, Family Noctuidae, Iroquois County.

Reared Cecropia Moth larva, Champaign County, showing crypsis.

Rearing Saturniids

The rearing of saturniids can be a rewarding educational experience. Much will be learned of the behavior and development of these magnificent insects. Fortunately, the species occur throughout the state, and none are listed as endangered or threatened. Breeding stock can be obtained by various means. Some species overwinter in cocoons that are conspicuous on bushes and trees during the winter. The easiest cocoons to find containing live pupae are those of the Cecropia Moth, the Promethea Moth, and the Polyphemus Moth. Overwintering pupae of other species, whether within a silken cocoon in the leaf litter below the larval host plant or buried in the soil without a cocoon, are more difficult to find. Such species are best obtained by collecting adults near a light at night during the summer months. A female moth collected at a light is almost certainly mated, and will lay viable fertilized eggs over several days. She should be placed in a paper bag or other receptacle. No plants are needed; the eggs will be attached to the paper bag. A female moth obtained from an overwintered cocoon presents a problem in that it must first mate before ovipositing, and if the resulting larvae are to be reared, emergence of the female moth must be synchronized with the spring and summer growth of the larval food plant. Keeping the cocoons outdoors in a protected place will ensure that development from pupa to adult occurs at the natural seasonal time for the species. Most species will mate under captive conditions by placing the two sexes in a screened cage or porch.

Larvae can be found by carefully searching the foliage of food plants. Look for feeding damage and for the large fecal droppings on the ground below the larva. To rear larvae, care must be taken to provide fresh food at all times. The stress caused by leaves of poor quality or the absence of food for too long will weaken the larvae and lead to disease. It is important that the larvae not be crowded. The best results are obtained by confining larvae in a net placed over an entire bush or tree, if not too large, or over a branch with abundant foliage. The fabric used for window screens in tents is suitable and will last for several years. The quantity of foliage needed by saturniid larvae is easy to underestimate. For example, an average-sized lilac bush will support less than 10, perhaps only 5, cecropia larvae to maturity. In addition to diseases, predators such as predaceous stink bugs and some carabid ground beetles will cause losses. Mice can also be a problem, tearing the netting to enter and then feeding at will on the larvae. Without protection, birds are apt to take all of the larvae. For more detailed information on rearing, we recommend referring to the literature cited in the Additional Reading section.

Photographing Insects

Successful photography of insects in nature requires proper equipment, and equally important, knowledge of the behavior of the insect and its probable reaction to the approach of the photographer and even the appearance of the equipment used. A single-lens reflex (SLR) camera will give the best results. The need for a large image of a small object (the insect) requires the use of a macro lens (one designed for close focus). For extremely small insects, special lenses and other devices may be needed for magnification; these will not be covered here. The reader is referred to texts covering this topic (see "Photography," page 90). But for those insects over a few millimeters in length, a lens system that will give up to a twice life-size image, or smaller in most cases, will be adequate. Most insects require an image ratio of 1:1 or less on the film. The longer the focal length of the lens, the less close the camera needs to be for a specific size of the image on the film. At the same time, the need for a tripod increases with the use of a longer focal length lens. For most insects, a tripod will be a handicap because of the loss of the photographer's mobility. Relatively few insects will be found sitting still, due either to air movement or to active motion by the insect itself. Very few insects will remain in place while the photographer sets up the tripod and camera. Very often it is necessary to follow the insect's image in the viewfinder by moving the camera as the insect moves. All of this means that either very sensitive film (high ASA/ISO) and an extremely rapid shutter speed must be used, or that electronic flash is used as the light source. With flash, slow film can be used (ASA/ISO 25 to 64) and yet allow the use of the smallest f-stops (f-16 or 22 usually), thus achieving the maximum possible depth of field and at the same time the best resolution because of the fine grain of the film. In addition, the extremely short duration of exposure with electronic flash (1/1000 second or less) stops motion of the subject, eliminating blurred images.

Exposure by flash has a disadvantage in that the background, if too far away, will be black. Thus, a flash-illuminated photograph may be less aesthetically pleasing than one taken under natural lighting.

Problems inherent with the photography of insects are due to insect behavior in response to the close presence of the photographer and his or her equipment. Many insects will be easily frightened and take evasive action. A quiet, slow, and stealthy approach is needed. The insect compound eye is superbly structured to discern movement. A slow approach lessens the chance of a reaction, that is, flight. No sudden moves should be made. Do not cast a moving shadow over the insect. Do not jingle keys or chains; the ultrasound produced may cause sudden flight. Not every insect will permit a close approach, but with patience some will be discovered that will. The photographer will then do best by taking a series of pictures. This increases the chances for a really good photograph. Remember that the cost of film is the least expensive part of the endeavor.

Insects that are most easily approached without sudden alarm are those that are feeding, whether on nectar from a flower, the juice from a fermenting apple, or the liquid on dung or carrion. Often an insect that has recently molted (more correctly—undergone ecdysis) will remain quiet even when disturbed. If the temperature is low, many insects will be less apt to take flight. The time of day may influence behavior, perhaps lessening the tendency to take flight.

Sometimes it may be necessary to first capture the insect, then anesthetize it by chemical or physical means. For some species, this may be the only way to get a picture. Chilling in a refrigerator is the safest and is tolerated by most species. To obtain a lifelike and natural-appearing image, the insect must be allowed to fully recover so that all appendages and the body will be in a natural position. Too many pictures have been published with raised tarsi projecting in the air. Once the insect has recovered, one or more pictures are usually possible. Chilling is especially productive with nocturnal insects, which can then be photographed the next day. In these cases, the chilled insect can be placed in a suitable site and photographed when it has recovered from the cold. Such insects generally remain quiet for extended periods.

Large insects, such as the nocturnal saturniids, when found during daylight hours, often (but not always) cannot fly without first increasing their body temperature to make flight possible. They do this by a quivering action of the thoracic muscles. During this warm-up there is ample time to obtain photographs without difficulty. Using this procedure, most of the photographs in nature of the saturniids in this book were taken of captive-reared insects.

Field Guide to Silkmoths of Illinois
Saturniid Relatives

Female *Apatalodes torrefacta* (Apatelodinae), Cook County.

Male *Apatalodes torrefacta* (Apatelodinae), Cook County.

Female Twin-spotted Sphinx Moth, *Smerinthus jamaicensis* (Sphingidae), Peoria County.

Male White-line Sphinx Moth, *Hyles lineata* (Sphingidae), Champaign County.

Species Accounts

Regal Moth or Royal Walnut Moth
Citheronia regalis (Fabricius, 1793)

Subfamily: Ceratocampinae

Note: The larvae are called Hickory Horned Devils because of their threatening appearance, although they are harmless.

Description of Adult: Wingspan 95–155 mm (3 3/4–6 1/8 in.). Females are larger than males. Sexes similar in color. In shape the wings are more elongate than those of our other saturniids. The body, especially the abdomen, is very stout. On the upperside, the forewing is olive-gray with veins lined with reddish brown scales. There are two basal and one discal yellow spots and a yellow-spotted postmedial band. The hindwing is reddish brown with a yellow anterior area, and distal patches of olive-gray. The body is reddish brown with yellow thoracic spots and circular bands between abdominal segments. On the underside the wings are yellow basally, and olive-gray with reddish brown veins on the distal areas. Yellow postmedial spots are present. At rest, the wings are flexed, held in a rooflike position over and along the sides of the abdomen (covered in "Classification" section).

Similar Species: None in Illinois.

Description of Larva: Up to 130 mm (5 1/4 in.) long, the largest caterpillar of our area. There are five instars; one and two are brown with long conspicuous spiny thoracic scoli, and shorter scoli on the abdominal segments. Instars three, four, and five are green, with short abdominal scoli. The thoracic scoli are stout and spiny, red with black tips. Those on the meso- and metathoracic segments are the longest. The thoracic legs are red; the abdominal prolegs are black and green. Each abdominal segment has a diagonal white bar with a black upper edge, shading into a dark brown area.

Description of Pupae: Pupae are smooth, with a stout cremaster. The species does not spin a cocoon; mature larvae dig down into the soil and form a cell within which to pupate. Pupae tunnel their way to the surface for eclosion of the adult by means of strong abdominal thrusts. In color they are like many moth pupae, that is, dark brown.

Habitat: Forests, woods, suburban areas, along roadsides where the larval food plants grow.

Natural History: Univoltine, although there are reports of a second generation in Missouri. Adults appear during the summer in Illinois. They often come to light, as do other royal moths. They do not feed, nor do they drink. The mouthparts, as is true of all saturniids, are nonfunctional. Old reports in the literature of feeding by *Citheronia* must be due to misidentification. Mating is nocturnal, in the hours before and after midnight. As with all saturniids, females release a pheromone creating an odor trail to their location that males can follow. Eggs are laid in small groups or singly. Upon hatching, the larvae disperse. The first two instars rest in a curled position, somewhat resembling a bird dropping. The later instars are green and brown and cryptic. When attacked, they attempt to defend themselves by violent movements of the head and thorax, swinging the elongate scoli from side to side. Larvae are polyphagous, found usually on hickories (*Carya* spp.), persimmon (*Diospyros virginiana*), sumac (*Rhus* spp.), sweet gum (*Liquidambar styraciflua*), and walnuts (*Juglans* spp.). Larvae can be found on other plant species occasionally, for example, cotton in the south.

Status: Common in the southern part of the state. Absent or scarce in the northern and west-central counties. In 50 years of collecting and observation, JGS has seen only one example in Urbana; it was by a light over an entrance to the Illini Union on the campus of the University of Illinois, Urbana-Champaign. In Pope County in southern Illinois, the species was (and is) common at light. *C. regalis* ranges from Massachusetts west through the Ohio River Valley, from south of Chicago to Oklahoma and eastern Texas, east to Florida and north to Massachusetts. It is common over much of this range, although disappearing in the northeast. It is normally found in heavily forested areas.

Reared Hickory Horned Devil (Regal Moth larva), early instar, Pope County.

Reared Hickory Horned Devil (Regal Moth larva), 5th instar, Pope County.

Regal Moth

Reared male Regal Moth, Pope County.

Reared female Regal Moth, Pope County.

Field Guide to Silkmoths of Illinois

Reared Regal Moth, Pope County.

Reared male Regal Moth, Pope County.

Regal Moth

Reared male Regal Moth, Pope County.

Reared male Regal Moth, Pope County.

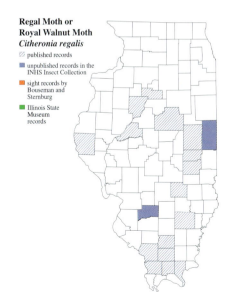

**Regal Moth or
Royal Walnut Moth**
Citheronia regalis

- published records
- unpublished records in the INHS Insect Collection
- sight records by Bouseman and Sternburg
- Illinois State Museum records

Field Guide to Silkmoths of Illinois

Pine-devil Moth   **Subfamily: Ceratocampinae**
Citheronia sepulchralis   Grote and Robertson, 1865

Note: The species is not a regular part of the Illinois fauna but has been collected in the state. We therefore include it in this guide book.

Description of Adult: Wingspan 70–100 mm (2 3/4–5 in.). Males much smaller than females. Sexes colored alike. Upperside of forewing olive-gray with faint rosy scales along the veins and a rosy spot at the wing base. A weak postmedial band is present without yellow spots. Basal region of the hindwing is rose colored; the outer regions are olive-brown with veins rosy. The body is entirely brown. On the underside, the wings are rosy.

Similar Species: None in Illinois.

Description of Larva: Length to 100 mm (4 in.). Brown with black shading not as colorful as *C. regalis*. The thoracic horns are yellow and shorter, a single pair on the second and on the third thoracic dorsum. Segments eight and nine of the abdomen each have a single dorsal scolus.

Description of Pupa: Similar to Regal Moth.

Habitat: Restricted to pine forests.

Natural History: A univoltine species. Sexes colored alike, but females may be twice the size of the males. The larvae feed only on pines, including *Pinus virginiana, P. rigida,* and *P. strobus*. These pines have been extensively planted in Illinois. The life history of the species does not differ significantly from that of *C. regalis.*

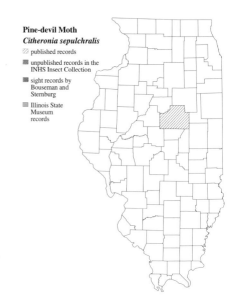

Pine-devil Moth
Citheronia sepulchralis
- published records
- unpublished records in the INHS Insect Collection
- sight records by Bouseman and Sternburg
- Illinois State Museum records

Status: Known only as a rare vagrant in Illinois. The species is locally common in the Atlantic states, from New England to Florida. It extends into eastern Ohio, Kentucky, and Tennessee, and could conceivably reach Illinois.

Pine-devil Moth

Male Pine-devil Moth, Buncombe County, NC.

Female Pine-devil Moth.

Imperial Moth
Eacles imperialis (Drury, 1773)

Subfamily: Ceratocampinae

Description: Wingspan 90–165 mm (3 1/2–6 1/2 in.). Sexually dimorphic, females mostly yellow, males with conspicuous magenta patches. Upperside yellow with magenta or purple-brown markings. The dark markings are variable in extent on the forewing submargin and the basal forewing and hindwing regions. A dark postmedial line crosses both wings; on the forewing it extends diagonally to the apex. Both wings have small circular gray-centered eyespots and a scattering of magenta dots. The body is yellow with magenta cross-bands. On the underside, both wings are yellow, with the eyespots magenta, and magenta patches reduced or absent.

Similar Species: None in Illinois. Related species occur in Arizona and south into the Neotropics. A northern subspecies of *E. imperialis* is found in upper Michigan and Ontario.

Description of Larva: Length of mature larvae may reach 115 mm (4 1/2 in.). Color may be green, brown, or black. Long hairlike setae are tan or white. The spiracles are white with dark blue edges. Meso- and metathoracic dorsal horns (scoli) are stout and spiny. Small scoli are present on the abdominal segments. The early instars have the scoli disproportionately long. Larvae are present from July through September.

Description of Pupa: The rather spinose brown pupa has a bifurcate cremaster. If disturbed, the abdomen moves in a violent twirling motion. The pupa overwinters in a cell formed by the larva before pupation; there is no cocoon.

Habitat: Forests, suburban areas, parks, roadsides. The wide variety of trees and shrubs that serve as larval hosts ensure wide distribution.

Natural History: Univoltine with a long extended emergence; adults present from June into August. Pupae are the overwintering stage. Adult emergence takes place near dawn, after which individuals remain quiet until dusk, at which time they become active. Both sexes are nocturnal. Males are strong fliers, soon finding females by following the odor trail of the sex pheromone released by a female, usually during the early morning hours after midnight. Eggs are laid singly or in small groups on foliage of larval hosts. The species is very polyphagous, larvae accepting many broad-leaved plants and even some conifers. A partial list includes birches (*Betula* spp.), maples (*Acer* spp.), oaks (*Quercus* spp.), pines (*Pinus* spp.), sassafras (*Sassafras albidum*), walnuts (*Juglans* spp.), and many others. Larvae disperse to feed and remain solitary throughout larval life.

Imperial Moth

Status: Common statewide. The species comes readily to light and is often seen at night. It ranges from Massachusetts to Iowa, south to Texas, and east to Florida and the Atlantic coast, and is often common, a denizen of deciduous forests and even urban areas.

Reared male Imperial Moth, Champaign County.

Reared female Imperial Moth, Champaign County.

Field Guide to Silkmoths of Illinois

Reared Imperial Moth, Champaign County.

Reared Imperial Moth, Champaign County.

Reared Imperial Moth larva (5th instar, brown form), Champaign County.

Reared Imperial Moth larva (5th instar, green form), Champaign County.

Imperial Moth

Reared Imperial Moth, Champaign County.

Reared Imperial Moth, Champaign County.

Imperial Moth pupa collected in 1932, Champaign County.

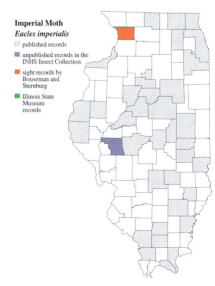

Imperial Moth
Eacles imperialis
- published records
- unpublished records in the INHS Insect Collection
- sight records by Bouseman and Sternburg
- Illinois State Museum records

Spiny Oakworm Moth
Anisota stigma (Fabricius, 1775)
Subfamily: Ceratocampinae

Description of Adult: Wingspan 40–70 mm (1 5/8–2 3/4 in.). Sexual dimorphism is not pronounced. Males are small and have the wings less rounded. The general color is bright orange-brown tinged with pink, and densely spotted with brown. The marginal area beyond the postmedial line has a purplish or pink cast in both sexes. The forewing has a round white discal spot, typical of *Anisota* spp. A postmedial line crosses the hindwing. At rest the wings are flexed.

Similar Species: Females of *A. senatoria* have less dense spots. Females of *A. virginiensis* do not have the forewings spotted with dark dots. Males of *A. senatoria* and *virginiensis* are smaller and hyaline, and diurnal, unlike the males of *A. stigma,* which are both nocturnal and diurnal.

Description of Larva: Larvae are called Spiny Oakworms. Length to 35 mm (1 3/4 in.). Head is orange-brown. The body is variable, from red to pink or brown, with a dense scattering of white granules. The second thoracic segment has a pair of long spiny horns. Dorsal and lateral short spinulose horns along the back and sides recurve backwards or downwards. There often is an indistinct spiracular stripe. Early instars are gregarious, the later instars solitary. These pupate in the ground. There is one generation in the north, two in the south. Larvae feed mainly on oak (*Quercus* spp.) and occasionally on hazel (*Corylus* spp.).

Description of Pupa: Typical of Ceratocampinae. Pupa with bifurcate cremaster. Body surface of thorax and abdomen spinose. Pupation in the ground with no cocoon.

Habitat: Forests and woodlots where there are oaks. Suburban areas, savannas.

Life History: Univoltine. Collecting at night with lights indicates that both sexes are nocturnal, but daytime observation reveals that males are also diurnal, actively searching for and mating with females even in broad daylight. Females lay their eggs after dark on oak leaves, the eggs in groups of 5 to 20 or so. Adult flight is spread over a long season in June and July, although the life span of any one individual is only 7–10 days.

Status: Uncommon, but widespread.

Spiny Oakworm Moth

Male Spiny Oakworm Moth, Cook County.

Female Spiny Oakworm Moth, Cook County.

Spiny Oakworm early instar larva. Photo by Phil Nixon.

Spiny Oakworm larva (late instar), Pope County.

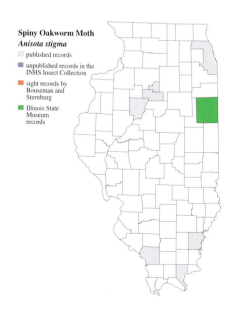

Pink-striped Oakworm Moth
Anisota virginiensis **(Drury, 1773)**

Subfamily: Ceratocampinae

Description of Adult: Wingspan of male 40–45 mm (1 9/16–1 3/4 in.), of female 50–65 mm (2–3 1/2 in.). Sexually dimorphic, females much larger than males. In the male the forewing has a well-developed hyaline area and dark brown borders. Its hindwing is dark with a weak postmedial line. The female has a well-developed postmedial line on both wings with purplish cast in the area beyond the lines. Both sexes have a well-developed white discal spot on each forewing. The wings do not have a scattering of dark brown spots. Adults have reduced mouthparts and neither feed nor drink.

Similar Species: *Syssphinx bicolor* and *S. bisecta* lack the hindwing postmedial line. *Anisota stigma, A. senatoria,* and *S. bicolor* have small dark spots on the wings.

Description of Larva: Length to 55 mm (2 3/8 in.). Known as the Pink-striped Oakworm. The head is orange-brown and the body is pale olive green, covered with white granules. The lateral areas are pink. On the mesothorax there is a pair of dorsal scoli, 6–8 mm long. The spiracles are black with white edges. Running the length of the body are wide subdorsal and spiracular white stripes.

Description of Pupa: Typical of Ceratocampinae. Pupa with bifurcate cremaster. Body surface of thorax and abdomen spinose. Pupation in the ground with no cocoon.

Habitat: Forests, savannas, and suburban areas where there are oaks (*Quercus* spp.).

Life History: Bivoltine. Adults in May and again in July to September. The species overwinters as pupae in the soil. Males are diurnal, with mating activity in mid to late morning. The males are fast and erratic fliers, but, when circling over a cage containing a virgin female, each resembles quite convincingly a wasp or bee. If the female is free within the cage and the male can enter, mating takes place at once. There is no courtship; this behavior is common to all of our saturniids. The pair stays together until dusk, when they separate and egg laying begins. Eggs are laid on oak leaves in groups of 100 or less. Larvae feed on the oak foliage, the early instars in groups, later instars separately.

Status: Statewide, often common. Occasional outbreaks cause considerable defoliation of infested trees.

Pink-striped Oakworm Moth

Male Pink-striped Oakworm Moth, Cook County.

Female Pink-striped Oakworm Moth, Cook County.

Pink-striped Oakworm Moth, Wayne County, MO.

Pink-striped Oakworm Moth, Wayne County, MO.

Pink-striped Oakworm larva. Photo by James Appleby.

Pink-striped Oakworm Moth
Anisota virginiensis
- published records
- unpublished records in the INHS Insect Collection
- sight records by Bouseman and Sternburg
- Illinois State Museum records

Orange-striped Oakworm Moth
Anisota senatoria (J.E. Smith, 1797)

Subfamily: Ceratocampinae

Description of Adult: Sexually dimorphic. Forewing of male is narrow with a translucent spot. Male wingspan 33–44 mm (1 1/4–1 3/4 in.). Female wingspan 50–65 mm (2–2 1/2 in.). Female has broad, opaque, rounded wings covered with small spots. In color males are reddish brown in the scaled areas; females are yellow-brown or reddish brown with pink shading beyond the postmedial line. Typical of *Anisota* spp., the forewing has a conspicuous white discal spot. Males are diurnal; females fly at night, but of course release pheromone to attract males during the day when mating occurs. Oviposition begins before dark and continues into the night.

Similar Species: Males of *A. virginiensis* have large transparent areas, while females lack spotting. Females of *A. stigma* are densely spotted and paler.

Description of Larva: The larva is known as the Orange-striped Oakworm. The body is black, as is the head. Eight (four on each half) light or yellow longitudinal stripes run the length of the body. There are two rows of black dorsal scoli and each side has one subdorsal row, all short except for the long pair on the mesothorax. Larvae are gregarious in the early instars, then dispersing, but often remaining close together. They feed on oaks and are periodically a major pest, causing serious defoliation of forest trees.

Description of Pupa: Typical of Ceratocampinae. Pupa with bifurcate cremaster. Body surface of thorax and abdomen spinose. Pupation in the ground with no cocoon.

Habitat: Forests where there are oaks (*Quercus* spp.).

Life History: Univoltine. Overwinters as pupae in the soil. Adults fly in late June and July. Males are diurnal, searching for females in late morning, when mating takes place. The pair separates in the late afternoon, after which the female begins to oviposit on oak foliage, continuing after dark. Larvae feed from late July into October.

Status: The species is uncommon, but occasional outbreaks cause serious damage to oaks. It is one of the few saturniids to attain pest status. The species ranges from Vermont to Minnesota to Texas and northern Georgia.

Orange-striped Oakworm Moth

Male Orange-striped Oakworm Moth, Cook County.

Female Orange-striped Oakworm Moth, Cook County.

Orange-striped Oakworm larva. Photo by James Appleby.

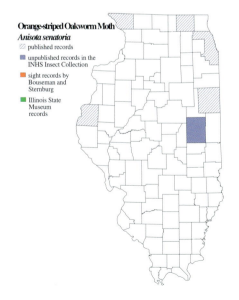

Orange-striped Oakworm Moth
Anisota senatoria
- published records
- unpublished records in the INHS Insect Collection
- sight records by Bouseman and Sternburg
- Illinois State Museum records

Rosy Maple Moth
Dryocampa rubicunda (Fabricius, 1793)

Subfamily: Ceratocampinae

Description of Adult: Wingspan 35–52 mm (1 3/8–2 1/8 in.). Wings on the upperside yellow to white with the basal area of the forewing and the outer margin bright pink. There may be a postmedial band of pink on the hindwing. The intensity of the pink varies from bright to very pale. The body is yellow. Some specimens are almost all white, with only a trace of pink.

Similar Species: *Hyparpax aurora* (J.E. Smith), the Pink Prominent Moth, family Notodontidae, is remarkably similar in its coloring but has a pink body. The noctuid *Schinia florida* (Guerin), the Primrose Moth, is similar except that the pink pattern differs. It has been suggested (Ferguson 1972, p. 11) that this may be a mimicry complex, with the Rosy Maple Moth the model. We know of no data proving this supposition, although it is certainly possible.

Description of Larva: The larva is known as the Green-striped Mapleworm. Full grown to 60 mm (2 3/8 in.). The last larval instar is pale green with seven longitudinal dark green stripes. The lateral areas of the seventh and eighth abdominal segments are pink to red. The mesothorax has a single pair of long black dorsal scoli. Short black hornlike lateral scoli are present from the mesothorax to the eighth abdominal segment. There is a pair of mid-dorsal scoli on the ninth abdominal segment. The entire body is covered with rounded granules. The head is orange-brown.

Description of Pupa: Typical of Ceratocampinae. Pupa with bifurcate cremaster. Body surface of thorax and abdomen spinose. Pupation in the ground with no cocoon.

Habitat: Forests where the larval hosts occur. Maples (*Acer* spp.), especially red (*A. rubrum*), silver (*A. saccharinum*), and sugar maples (*A. saccharum*), and sometimes oaks (*Quercus* spp.) are the food plants favored.

Natural History: Populations are univoltine in the north, multivoltine in the south where there can be three generations. Pupae of the last generation overwinter in cells in the soil made by the last generation larvae. First generation adults appear in May. They come to light readily and are nocturnal, resting quietly during daylight hours. Like all saturniids the adults neither feed nor drink, and depend for energy entirely on the reserves stored during the larval stages. Adults survive for 1 week to 10 days. First instars are gregarious; later instars are solitary.

Rosy Maple Moth

Status: Usually common, sometimes abundant. It can be found statewide.

Remarks: At rest, the Rosy Maple Moth holds its wings in a flexed position over and against the sides and upper surface of the abdomen.

Male Rosy Maple Moth (light form), Cook County.

Male Rosy Maple Moth (pink form), North Carolina.

Green-striped Maple Worm (late instar). Photo by Phil Nixon.

Male Rosy Maple Moth (pink form), Champaign County.

Rosy Maple Moth (light form), Wayne County, MO.

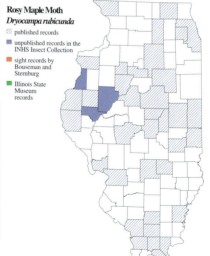

Rosy Maple Moth
Dryocampa rubicunda
published records
unpublished records in the INHS Insect Collection
sight records by Bouseman and Sternburg
Illinois State Museum records

Rosy Maple Moth (pink form), Wayne County, MO.

Bisected Honey Locust Moth
Syssphinx bisecta **(Lintner, 1879)**

Subfamily: Ceratocampinae

Description of Adult: Wingspan 55–75 mm (2 1/8–3 in.). Sexes similar, with males smaller. General color varies from yellowish to orange. A reddish area of the hindwing varies in size. The variations are not seasonal, but occur in each generation. The forewings of females are stippled with dark spots, usually absent in males. A thin, black postmedial line runs from the forewing apex to the base of the inner margin, bisecting the forewing.

Similar Species: None. The thin black postmedial line is diagnostic for *S. bisecta*.

Description of Larva: Length to 50 mm (2 in.). Body green with silvery white granules. Second and third thoracic segments each with one pair of long orange to blue hornlike scoli. The caudal horn is yellow at its base, red to black towards the tip. Conspicuous abdominal tubercles are conical and sharp, shiny white, and directed upward and backward. The spiracular stripe is blue above and yellow below.

Description of Pupa: Typical of the royal moths, spinulose, with a prominent cremaster. Pupation is in the soil, with no cocoon.

Habitat: Forests where there are honey locusts and/or Kentucky coffee trees.

Life History: Overwinters as a pupa without a cocoon in the soil. The species is bivoltine in the Midwest, with adults in May and July. Emergence of first generation adults is in May, the second generation appearing in July. There is apparently little if any seasonal dimorphism. Emergence is in the late afternoon, with mating occurring from late evening to early morning. The female begins oviposition the next night on honey locust (*Gleditsia triacanthos*) and Kentucky coffee tree (*Gymnocladus dioicus*). *S. bisecta* and *S. bicolor* are sympatric and active at the same time. Hybridization is apparently prevented by differences in the pheromones released by the females.

Status: Not uncommon. Probably statewide, but usually less common than *S. bicolor*.

Remarks: The generic name *Sphingicampa* is used by some authorities.

Bisected Honey Locust Moth

Male Bisected Honey Locust Moth, Champaign County.

Female Bisected Honey Locust Moth, Peoria County.

Male Bisected Honey Locust Moth, Champaign County.

Bisected Honey Locust Moth larva. Photo by James Appleby.

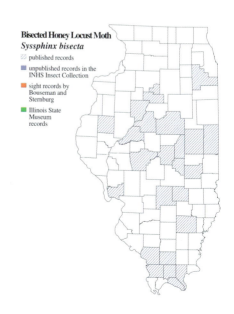

Bisected Honey Locust Moth
Syssphinx bisecta
- published records
- unpublished records in the INHS Insect Collection
- sight records by Bouseman and Sternburg
- Illinois State Museum records

Field Guide to Silkmoths of Illinois

Honey Locust Moth
Subfamily: Ceratocampinae
Syssphinx bicolor (Harris, 1841)

Description of Adult: Wingspan 45–65 mm (1 7/8–2 5/8 in.). Sexually dimorphic, seasonally polymorphic. Males are much smaller than the females and have less-rounded wings. In all seasonal forms the forewings are colored for concealment; the hindwings are colored with red that is revealed when the moth is disturbed (see photographs, page 40). Forewings of the spring brood are gray, yellow-orange in the summer brood, and brown in the third brood. An indistinct postmedial line reaches the forewing costa before the apex. A light to heavy dusting with brown spots occurs on the forewings. The white discal spot of the forewing may be double, single, or absent.

Similar Species: *Anisota* spp. lack a red patch on the hindwings, which have a distinct postmedial line. *Syssphinx bisecta* has the postmedial line of the forewing extended to its apex.

Description of Larva: Length to 55 mm (2 1/8 in.). The body is green with a bluish cast. The lower side of the abdomen bears scattered silvery white and black granules. There is a spiracular abdominal line, reddish purple above and white below. The second and third thoracic segments both have a pair of hornlike scoli, rose to red in color, and tipped with black. They are recurved in position. The abdominal segments have a variable number of short dorsal and subdorsal horns, silver and bright pink in color. There is a prominent mid-dorsal silver and coral-red horn on the dorsum of the eighth abdominal segment.

Habitat: Forests where there are honey locusts and/or Kentucky coffee trees.

Description of Pupa: Typical of the subfamily, spinose with a well-developed cremaster.

Life History: Pupae of the last generation overwinter in cellular cavities formed in the ground by larvae. Emergence of the first brood of adults is in May. Their larvae (and all subsequent larvae) feed on the leaves of honey locust (*Gleditsia triacanthos*) and Kentucky coffee tree (*Gymnocladus dioicus*), then pupate in the soil. Second-brood adults are found in midsummer, July and August; third brood, if present, appears in September. Adult emergence occurs in the evening, followed by mating that night. The next evening the pair separate and the female begins oviposition that night. The male will mate again if it can find another female. The female mates only once. Eggs are laid singly or in small groups on leaves of the host plant. Hatching occurs within five days or more. Larvae develop rapidly, becoming full grown within three weeks. The pupal stage in the ground lasts two weeks, except for the overwintering pupae that have entered diapause.

Honey Locust Moth

Status: Common to scarce, depending on local conditions. Statewide.

Remarks: The generic name *Sphingicampa* is used by some authorities.

Male Honey Locust Moth (spring form), Champaign County.

Female Honey Locust Moth (spring form), Champaign County.

Male Honey Locust Moth (summer form), Champaign County.

Female Honey Locust Moth (summer form), Champaign County.

Honey Locust Moth larva (5th instar), Champaign County.

Honey Locust Moth (resting position, spring form), Champaign County.

Field Guide to Silkmoths of Illinois

Honey Locust Moth (defensive display, spring form), Champaign County.

Honey Locust Moth (defensive display, summer form), Champaign County.

Honey Locust Moth (resting position, summer form), Champaign County.

Honey Locust Moth
Syssphinx bicolor

- published records
- unpublished records in the INHS Insect Collection
- sight records by Bouseman and Sternburg
- Illinois State Museum records

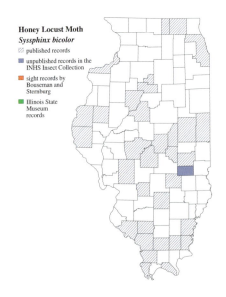

Buck Moth
Southern Illinois Population
Hemileuca maia maia **(Drury, 1773)**

Subfamily: Hemileucinae

Description of Adult: Wingspan 50–75 mm (2–3 in.). Sexes similar, except the abdomen is entirely black in the female and males have the tip of the abdomen bright red. Both wings are black, each crossed by a white medial band with a black discal spot that touches the black basal area. The undersides of the wings are much like the uppersides. Male antennae are bipectinate with long rami almost to their tip. Female antennae are bipectinate with short rami.

Similar Species: *H. nevadensis* of the Great Lakes complex. See discussion under that species.

Description of Larvae: Length to 60 mm (2 3/8 in.). The color varies from off-white to black, dusted with yellow specks. There may be an indistinct white spiracular stripe. The body is covered with dense tufts of urticating spines, a deterrent against predators. Handle with care! The toxic substance released by the spines may cause a burning sensation or even swelling. Larvae feed on oaks (*Quercus* spp.). There are six larval instars.

Description of Pupa: Typical of *Hemileuca* spp., there is a short simple cremaster with hooks. The abdominal segments do not telescope and the body is without exposed spines. The pupa is enclosed in a cocoon beneath leaf litter, where it is spun by the larva.

Habitat: Oak forests.

Life History: Univoltine. Overwinters as masses of eggs encircling twigs of the larval host, or sometimes on adjacent nonhost plants. Eggs are laid in late October. After the oaks leaf out in the spring, the eggs hatch. The first few instars remain together, feeding on the leaves; later instars tend to disperse, and during this movement may wander onto plants that are not fed upon. Because of this, reports of feeding on a variety of plant species that are not food plants may be in error. One that does appear to be correct is hazel (*Corylus* spp.). Larvae pupate on or in the ground where there is leaf cover in late August and September. Emergence of adults is in October at the time of leaf fall. Both sexes are diurnal, with males searching for and mating with the females in bright sunlight during the late morning and early afternoon hours. On separating, egg-laying begins that day. Buck Moths are swift fliers, not easily captured. Resting adults when disturbed often fall to the ground with wings folded together above the thorax and the abdomen curled and extended so that the red tip (in males) is conspicuous. In this way it increases the aposematic appearance in much the same manner as the unrelated tiger moths such as the Salt Marsh Caterpillar Moth. The display lasts a few minutes or until the disturbance is over before the moth climbs back up a support. There is evidence that some pupae delay emergence until their second summer.

Status: Local to uncommon in southern Illinois, with a few records from central Illinois that may not be accurate. Buck Moths resembling *H. m. maia* occur in northeast Illinois, but feed only on willows (*Salix* spp.) or poplar (*Populus* spp.). They are thought to be *H. nevadensis*, or a hybrid of that species with *H. m. maia*.

Male Buck Moth, Hardin County (James Wiker Collection).

Male Buck Moth, Hardin County (James Wiker Collection).

Buck Moth larva (late instar), Franklin County, MA. Photo by James Tuttle.

Female Buck Moth ovipositing, Vinton County, OH. Photo by James Tuttle.

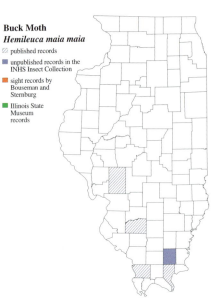

Buck Moth
Hemileuca maia maia
published records
unpublished records in the INHS Insect Collection
sight records by Bouseman and Sternburg
Illinois State Museum records

Nevada Buck Moth
Hemileuca nevadensis **Stretch, 1872**

Subfamily: Hemileucinae

Great Lakes Population: In the northeastern counties of Illinois there is a population of buck moths that resembles *H. maia maia* of the eastern and southern U.S., but whose larvae feed on *Salix* spp. (willows) and sometimes *Populus* spp. (poplars), and not on *Quercus ilicifolia* (scrub oak) or other oaks. This willow-feeding population extends from Minnesota through Wisconsin and Michigan to Illinois, Indiana, and Ohio. In the southern part of this range, from Illinois to Ohio, the adults are essentially indistinguishable from *H. m. maia*, and identifiable in a collection only by their locality labels. From the dark "*maia*-like" population northward and westward there is a cline leading to a lighter and more translucent population that merges in Minnesota with typical *H. nevadensis*. The latter ranges from Minnesota west to the Pacific Coast, always feeding on willows and poplars. Through the Great Lakes region, willows are the usual host, although some local populations are found on bogbean (*Menyanthes trifoliata*), dwarf birch (*Betula pumila*), and purple loosestrife (*Lythrum salicaria*). Apparently, none are on oak. In Ohio, the Great Lakes willow-feeding population meets the true oak-feeding *H. maia maia*. In Illinois the dark "*maia*like" population survives on willow or poplar. The distribution of the two taxa suggests hybridization may have occurred, but no data are available for substantiation. Another possibility is mimicry, perhaps Müllerian, with the western *H. nevadensis* evolving a color pattern like that of the aposematically colored *H. maia maia*.

Description of Typical *H. nevadensis*: Wingspan 50–70 mm (2–2 3/4 in.). From Minnesota west, typical Nevada Buck Moths have less-intense black basal and marginal areas than *H. m. maia*. The white medial area is broader, and the forewing discal spot does not touch the dark basal region. The wings appear more translucent than those of *H. m. maia*. The male has a red-tipped abdomen; the female has an all-black abdomen.

Description of Illinois *H. nevadensis*: The adult moths are identical in appearance with *H. m. maia*, but differ greatly in choice of larval host plant.

Description of Larvae: Not unlike those of *H. m. maia*.

Description of Pupae: Similar to *H. m. maia*.

Habitat: Wetlands where willows grow.

Life History: The species is univoltine. Eggs are laid in bands encircling the

Field Guide to Silkmoths of Illinois

twigs of willow (or sometimes a nearby plant). These hatch in the spring after the willows leaf out. The Nevada Buck Moth of northern Illinois is known to feed on willow, sometimes on poplar, but not on oak. In late summer they pupate in silken cocoons in the leaf litter on the ground. Emergence is in October. Adults are diurnal, mating and ovipositing during bright sunlight. The dark males behave much like aposematic tiger moths when disturbed, falling to the ground with wings together above the thorax and the abdomen curled and extended. This defense is typical of many unpalatable moths.

Status: Because of loss of wetland habitat, the moth is more local in distribution, but uncommon to common where it occurs in the northern sector of the state.

Male Nevada Buck Moth (Great Lakes population), Hessville, IN.

Female Nevada Buck Moth (Great Lakes population), Hessville, IN.

Male Nevada Buck Moth (typical of western population), California.

Nevada Buck Moth larva (late instar), Polk County, MN. Photo by James Tuttle.

Larval aggregation (early instar) with parasitoid on left, Lucas County, OH. Photo by James Tuttle.

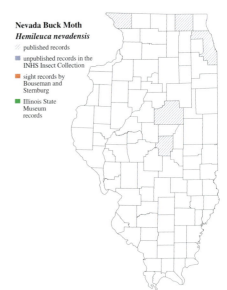

Nevada Buck Moth
Hemileuca nevadensis
published records
unpublished records in the INHS Insect Collection
sight records by Bouseman and Sternburg
Illinois State Museum records

Io Moth
Automeris io (Fabricius, 1775)

Subfamily: Hemileucinae

Description of Adult: Wingspan 50–85 mm (2–3 1/4 in.). Sexually dimorphic. Males much smaller than females. On the upperside males are mostly yellow with scattered brown spots and a weak broken-spotted medial line nearly parallel with the outer margin of the forewing. In contrast, the forewings of the female are purplish red to reddish brown instead of yellow and are otherwise similar. The hindwings of the two sexes are similar, each wing with a black discal spot with a blue and white center, a reddish brown postmedial band partially surrounding the eyespot, and a submarginal band beyond, with a yellow wing margin. The inner margin of the hindwing is red. On the underside the male is yellow, the postmedial lines are red, and on the hindwing the inner margin is red. The eyespots are black with white centers. The female is red-brown instead of yellow on the underside and is otherwise similar to the male. At rest the wings are held flexed over the abdomen, concealing the hindwing eyespots. Males have the antennae quadripectinate nearly to the tip; the females have bipectinate antennae.

Similar Species: None in Illinois. Related species occur in the western United States and south into the Neotropics.

Description of Larva: Length to 60 mm (2 3/4 in.). Rather stout green body with dense tufts of green spines, finely branched into urticating (stinging) spinules. These can cause a painful rash and swelling if touched. Handle with great care! The spiracles are white. A conspicuous red lateral line is present, bordered below by white and passing through the spiracles. Thoracic and abdominal legs are red. There are five larval instars.

Cocoon: The species overwinters as a pupa in a somewhat shapeless but sturdy cocoon spun among fallen leaves and ground litter.

Habitat: Forests, savannas, woody roadsides, occasionally in cornfields. Widespread.

Natural History: Univoltine in the north, bivoltine in Illinois. Overwinters as a pupa in a cocoon on the ground. Emergence is in May, and adults of the second generation are found throughout the summer. Look for them at night at lights. The species is extremely polyphagous, the larvae feeding upon most species of deciduous trees and shrubs, and on coarse grasses including corn (*Zea mays*). They have also been found on clover and other forbs. The list of trees includes apple (*Malus* spp.), basswood (*Tilia* spp.), blackberry (*Rubus* spp.), black cherry (*Prunus serotina*), elm (*Ulmus* spp.), hackberry (*Celtis* spp.), hickory (*Carya* spp.), maple (*Acer* spp.), oak (*Quercus* spp.), poplar (*Populus* spp.), privet (*Ligustrum* spp.), redbud (*Cercis canadensis*), sassafras (*Sassafras albidum*), willow (*Salix* spp.), and others. First and second instars are gregarious, brown in color; the later instars are green and solitary. All have stinging spines.

Status: Usually common, sometimes abundant. Several years ago willows along Route 150 near Oakwood were almost completely defoliated by this moth. The species probably occurs in every county in the state. It is widespread throughout eastern North America. The species ranges from Nova Scotia to Manitoba, south to Texas and Florida.

Reared male Io Moth, Champaign County.

Io Moth cocoon.

Reared female Io Moth, Champaign County.

Reared Io Moth larva (5th instar), Champaign County.

Io Moth

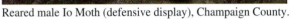

Reared male Io Moth (resting position), Champaign County.

Reared male Io Moth (defensive display), Champaign County.

Reared female Io Moth (defensive display), Champaign County.

Reared female Io Moth, (resting position) Champaign County.

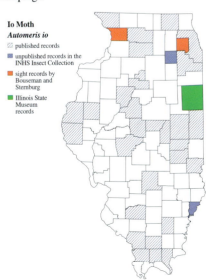

Io Moth
Automeris io
- published records
- unpublished records in the INHS Insect Collection
- sight records by Bouseman and Sternburg
- Illinois State Museum records

Polyphemus Moth
Antheraea polyphemus (Cramer, 1776)
Subfamily: Saturniinae

Description: Wingspan 110–150 mm (4 1/4–6 in.). Sexes similar, female forewing apex rounder than in male. General color varies from light brown or tan to yellowish brown, reddish brown, orange, or even melanic. On the upperside each wing has a transparent discal eyespot, the largest on the hindwing. The eyespot of the forewing is narrowly bordered with yellow and black. On the hindwing the eyespot has a narrow yellow partial border and a broad asymmetrical black border. The hindwing eyespots give the impression of a pair of shiny vertebrate eyes. There is a black submarginal line on both wings, lined outwardly with pink. The forewing has a broken basal dark line, basally bordered by pink. On the underside the wings have a strong resemblance to a dead leaf, with a scattering of light tan and dark brown areas. The eyespots are present, visible as transparent spots, but without the bordering colors that give the upperside spots a resemblance to vertebrate eyes. Thus, at rest, with the wings held vertically and together, the moth is cryptic, blending in with dead leaves. When the moth is first disturbed the wings are held outstretched, suddenly revealing the upperside eyespots. Known as a flash coloring display, this may deter some predators, although certainly not the Blue Jays in the backyard of one of us (JGS). Further disturbance causes the moth to bounce along the ground by downward thrusts of the forewing and elevating the hindwings at each bounce. This action further deters some predators that are frightened off by the staring (although false) eyes. Antennae of the males are broadly quadripectinate; those of the female are narrowly quadripectinate.

Similar Species: None in Illinois. Related species occur in Asia.

Description of Larva: The five larval instars are all bright green with an oblique yellow line on each side of abdominal segments two through nine. Tubercles (scoli) are small and reddish with a metallic iridescence. In silhouette the larval segments are convex, giving the appearance of a jagged leaf. The last larval instar is 75 mm (3 in.) or more in length.

Description of Cocoon: The egg-shaped cocoon is spun within leaves on the larval host plant and fastened in place by silken threads along the leaf stem. The connection is not strong; many cocoons fall to the ground during or before winter. The cocoon is thick-walled, stiff and hard, and has no escape valve. As with the Luna Moth, the eclosing adult softens the silk, then tears the weakened threads by means of hard spurs at the bases of the forewings.

Habitat: Forests and urban areas.

Polyphemus Moth

Natural History: Bivoltine in Illinois. Adults of the first generation from overwintered pupae emerge in May and June. Their progeny become adults in July and August. First-generation adults have wingspans of up to six inches or more. Second-generation adults are almost always smaller, irrespective of the plant species consumed by the larvae. Eclosion of the adults is usually in the afternoon. Mating takes place that night sometime near midnight or later. The pair stays together until the next evening. Females lay the white and brown eggs singly or in small groups on the larval host plant. Upon hatching (in 10 days or so), the larvae disperse to feed. The species is extremely polyphagous; more than 50 kinds of trees or shrubs serve as hosts. These include apple (*Malus* spp.), birches (*Betula* spp.), cherries (*Prunus* spp.), dogwoods (*Cornus* spp.), elms (*Ulmus* spp.), maples (*Acer* spp.), oaks (*Quercus* spp.), poplars (*Populus* spp.), plums (*Prunus* spp,), roses (*Rosa* spp.), walnuts (*Juglans* spp.), willows (*Salix* spp.), and many others. One of us (JGS) once found a large patch of wild roses in a vacant Chicago lot with over 100 cocoons present.

As is true of all saturniids, the adults do not feed or drink. They survive only as long as their energy reserves last, usually in less than a week to 10 days. Females release a pheromone to attract males. A female mates with the first male to find her and never mates again.

Status: Common statewide. It is often most common in rural areas. The practice of raking and disposing of leaves in urban areas must cause destruction of cocoons on the ground. The species thrives along roadsides and in wooded areas in general. The Polyphemus Moth is found from the Atlantic to the Pacific, and from Canada into Mexico, and is locally common throughout its range.

Reared male Polyphemus Moth, Champaign County.

Field Guide to Silkmoths of Illinois

Reared female Polyphemus Moth, Champaign County.

Reared male Polyphemus Moth (reddish brown form), Champaign County.

Polyphemus Moth

Reared Polyphemus Moth larva (5th instar), Champaign County.

Polyphemus Moth recently spun cocoon, Champaign County.

Overwintering Polyphemus Moth cocoon, Champaign County.

Polyphemus Moth pupa, Champaign County.

Reared male Polyphemus Moth (light-brown form), Champaign County.

Field Guide to Silkmoths of Illinois

Reared Male Polyphemus Moth (resting position), Champaign County.

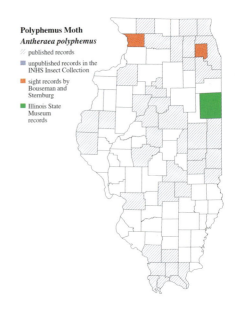

Polyphemus Moth
Antheraea polyphemus
▨ published records
■ unpublished records in the INHS Insect Collection
■ sight records by Bouseman and Sternburg
■ Illinois State Museum records

Luna Moth
Actias luna (Linnaeus, 1758)

Subfamily: Saturniinae

Description: Wingspan 95–135 mm (3 3/4–5 1/4 in.). Sexes similar, although the hindwing tails of the females tend to be shorter. Seasonally dimorphic; first generation with wing margins red, second generation with wing margins yellow. On the upperside, the costal margin of the forewing has a conspicuous broad brown edging. Fore and hindwings both have a discal eyespot, edged with black and white with a transparent center. The eyespot of the forewing connects to the costal band by a red-brown bar. The overall color is pale green, with white, or very pale yellow body setae. A prominent character is the greatly extended inner margin of each hindwing forming two long green tails. The undersides of the wings are pale green, with the eyespots faintly visible.

Similar Species: None in Illinois. Related species are Eurasian.

Description of Larva: All five instars are green with prominent segmentation. The last instar attains a length of about 75 mm (3 in.). On each side there are a dorsal and two lateral rows of small red scoli all covered with short setae. A yellow line passes through the spiracles. There may also be narrow vertical yellow bands between the abdominal segments. When fully fed, the last (fifth) instar changes color from green to reddish brown. At this time the gut is emptied of its contents.

Description of Cocoon: Pupation is within a silk cocoon spun by the fifth instar. The cocoons are spun among the fallen leaves and other litter on the ground, where they hide and are protected by the accumulated ground cover. In spinning its cocoon, the larva first pulls the edges of a leaf together, forming a shelter within which a thin-walled silken cocoon is formed without an escape valve. Emergence by the adult is facilitated by its softening of the silk by means of a regurgitated fluid, after which the weakened silk at the anterior end of the cocoon is torn apart by means of small sharp thoracic spurs.

Habitat: Forests, although sometimes wooded urban areas when walnuts and hickories, the principal larval hosts in Illinois, are present.

Natural History: Bivoltine in Illinois. Overwintering pupae complete development as warm temperatures return, emerging as first-generation adults in May and June. Their progeny become adults in late July and August. A partial third generation sometimes occurs in the south. The species is polyphagous. Females lay their eggs singly or in small groups on a variety of larval host plants. In Illinois hickories (*Carya* spp.) and walnuts (*Juglans* spp.) are favored. Other hosts include sumacs (*Rhus* spp.), persimmon (*Diospyros virginiana*), sweet gum (*Liquidambar styraciflua*), and occasionally others. White birch (*Betula papyrifera*) is used by more-northern populations.

Luna Moths after emergence remain quiet and inactive until evening, when flight begins. Mating usually takes place several hours after nightfall, often after midnight. Once coupled, the pair will remain together until dusk of the following day, unless disturbed earlier. Transfer of sperm is complete within 15 minutes of copulation. Males locate females by flying upwind toward the source of the odor trail created by a virgin female as she releases the sexual pheromone. Usually the first male to locate the female mates with her without any preliminary courtship. Once mated, female saturniids never again release pheromone; they mate only once. Males, however, can mate each night of their short life, only a week to 10 days in duration. During this time the females lay their eggs singly or in groups on the foliage of the larval host plant. The eggs hatch in about one week. The larvae are not gregarious, but disperse to feed.

Status: Common throughout Illinois. The species occurs in the eastern U.S. and lower Canada from the eastern Great Plains to the Atlantic Coast and from Nova Scotia to Florida and Ontario to Texas. It is often the most common wild silkmoth present.

Remarks: A Luna Moth in flight is an impressive sight, especially if it is seen during the day when the moth is frightened into flight. The movements of the wings with the inner margins of the hindwings held close to the abdomen result in the long trailing tails twisting in a rolling motion against one another, a motion that adds greatly to the moth's beauty and charm. The unusual wing shape and beauty of the Luna Moth have contributed to its popularity with amateur collectors and naturalists everywhere. Once seen, it is never forgotten—a truly spectacular insect. The Luna Moth has also been called the Moon Moth.

Reared Luna Moth larva (5th instar), Champaign County.

Reared Luna Moth cocoon, Champaign County.

Luna Moth

Reared male Luna Moth (spring form), Champaign County

Reared male Luna Moth (summer form), Champaign County.

Field Guide to Silkmoths of Illinois

Reared male Luna Moth (startled), Champaign County.

Reared male Luna Moth (resting position), Champaign County.

Luna Moth

Reared female Luna Moth (resting position), Champaign County.

Luna Moth pupa.

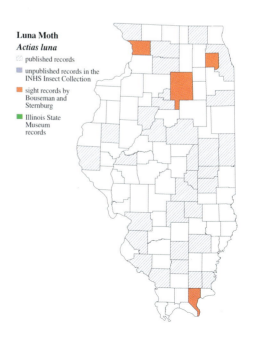

Luna Moth
Actias luna

- published records
- unpublished records in the INHS Insect Collection
- sight records by Bouseman and Sternburg
- Illinois State Museum records

Promethea Moth or Spicebush Silkmoth
Callosamia promethea (Drury, 1773)
Subfamily: Saturniinae

Description of Adult: Wingspan 75–95 mm (3–3 3/4 in.). Sexually dimorphic. Male black on the upperside with sinuous postmedial whitish line across both wings, a small forewing apical eyespot, and a violet apical area. The wing margins are light buff. An occasional individual may have faint discal spots. On the underside the basal areas are purplish black with a postmedial separation from a plum-colored region. Females have the same wing patterns, along with angular discal spots, but the general color is yellow-brown to red-brown and not black. The female wings are more rounded than those of the male.

Similar Species: Females of *C. angulifera* lack the reddish color of *C. promethea* females, and have the discal spots larger. Males of *C. promethea* in active flight can be mistaken for a dark blackish swallowtail butterfly. Experimental studies have shown that they are Batesian mimics of the toxic and unpalatable Pipevine Swallowtail (*Battus philenor*) in flight. See discussion below.

Description of Larva: Length to 75 mm (3 in.). Body is light green with a slight bluish cast. Second and third thoracic segments bear paired bright red scoli. There is a single yellow to red scolus on the dorsum of the eighth abdominal segment. Other segments bear small black reduced scoli. There is no low lateral white band, as in *C. angulifera* larvae.

Description of Pupa and Cocoon: The smooth brown pupa has no cremaster; it is enclosed within a double-walled cocoon spun by the fifth (last) larval instar within a rolled leaf, usually of the larva's host plant. Silk spun by the larva along the leaf stem and adjacent twig prevents the cocoon from dropping to the ground. They are therefore easily found hanging from the branches in winter when the trees are bare.

Habitat: Promethea is most successful where trees are somewhat isolated, as along fence rows or the edges of fields, in savannas, and in suburban or urban areas where the trees are not too numerous. As a rule, the species does not do

Reared Promethea Moth larva (5th instar, side view), Champaign County.

Reared Promethea Moth larva (5th instar, dorsal view), Champaign County.

well in heavily forested sites. Larvae are polyphagous, feeding on many plant families. In Illinois we have found sassafras (*Sassafras albidum*), tulip tree (*Liriodendron tulipifera*), and wild black cherry (*Prunus serotina*) to be favored. Other common larval hosts are ash (*Fraxinus* spp.), lilac (*Syringa* spp.), maples (*Acer* spp.), spicebush (*Lindera* spp.), sumac (*Rhus* spp.), sweet gum (*Liquidambar styraciflua*), and others.

Life History: In the north, Promethea is univoltine, with adult emergence in June. In central and southern Illinois the species is bivoltine, with the first generation appearing in May and early June from overwintered pupae, and the second-generation adults present in August. In a bivoltine population, pupal progeny from the first-generation adults do not diapause but develop directly to the second-generation adults. All the progeny of the second generation enter pupal diapause, as do a few from the first generation, and must undergo winter conditions to develop further.

Male Promethea Moths are diurnal, active from early afternoon to late afternoon, during which time they actively search for females by following the odor trails of pheromone released by calling females. Upon successful location of a female, mating takes place without courtship, at which time the female ceases to release pheromone. The couple stays together until evening when the female breaks away and begins oviposition activity. Males are inactive at night; they do not come to light. Females, however, sometimes do come to light. Females do not fly during daylight hours but of course are active in releasing pheromone at the proper time. As with Saturniidae in general, females mate only once.

In flight, the black Promethea Moth males resemble black-colored swallowtail butterflies. The flight speed and wing-beat frequency, to our eye, appears similar, and it is often difficult to recognize that the moth is not a butterfly. Experimental evidence indicates that the species is a Batesian mimic of the Pipevine Swallowtail (*Battus philenor*), which is known to be unpalatable or even toxic when eaten by a bird. Laboratory tests have shown that Promethea Moths are palatable to birds. Males fly from early afternoon to near dusk; much of this time coincides with the flight activity of Pipevine Swallowtails. In field tests, male Promethea Moths were painted to resemble either the palatable edible yellow Tiger Swallowtail (*Papilio glaucus*) or the unpalatable Pipevine Swallowtail. These were released in equal numbers in an area with mature forest, prairie, and second growth on former agricultural fields. The released painted male moths were reassembled by attraction to traps baited with virgin female moths. The results, here briefly summarized from data obtained over three summers, showed that moths painted to resemble an unpalatable butterfly survived longer and with less bird damage after release than those painted to resemble a palatable butterfly. Many of the recaptured yellow-painted moths (the Tiger Swallowtail mimics) had clear evidence of attack by birds, that is, beak-shaped tears on the wings. Damage to the black-painted moths (the Pipevine Swallowtail mimics) was comparatively minimal.

Status: Common to uncommon locally. Statewide distribution.

Remarks: One of us (JGS) remembers his first encounter with the diurnal Promethea males coming to newly emerged female Promethea Moths held within a screened porch in Chicago, near Evergreen Park. Two females had eclosed that morning and began to call (release pheromone) that afternoon. The air became crowded with males seeking females. A conservative estimate of their numbers was near 50. It was an impressive sight, especially to him at the age of 14.

Male Promethea Moth, Coles County.

Reared Promethea Moth larvae (early instar), Champaign County.

Reared Promethea Moth larvae (3rd instar), Champaign County.

Reared Promethea Moth egg mass, Champaign County.

Promethea Moth

Reared female Promethea Moth, Champaign County.

Reared Promethea Moth mated pair (female upper), Champaign County.

Reared Promethea Moth cocoon, Champaign County.

Promethea Moth pupa, Champaign County.

Reared male Promethea Moth recently emerged on cocoon, Champaign County.

Reared male Promethea Moth (resting postion), Champaign County.

Promethea Moth

Reared female Promethea Moth (disturbed position), Champaign County.

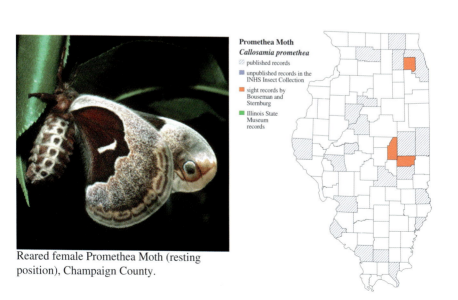

Reared female Promethea Moth (resting position), Champaign County.

Promethea Moth
Callosamia promethea
- published records
- unpublished records in the INHS Insect Collection
- sight records by Bouseman and Sternburg
- Illinois State Museum records

Tulip Tree Silkmoth
Callosamia angulifera (Walker, 1885)

Subfamily: Saturniinae

Description of Adult: Wingspan 80–110 mm (3 1/8–4 3/8 in.). Sexually dimorphic, males darker than females. Seasonally dimorphic, the summer brood males darker than those of the spring brood. The two sexes are alike in the pattern of the markings but differ in coloring. Males are light brown to dark brown, the females yellow-brown to orange-brown. Both wings are crossed by an irregularly waved, white postmedial line. Angulate discal spots are present on the wings, the largest on the forewings. The area basad to the postmedial line is dark, whereas the area beyond the line is light-shaded. On the underside of the male the area beyond the postmedial line is light pink, contrasting greatly with the dark brown basal area. In the female, differences in contrast are similar to those of the male, but the dark areas are reddish brown.

Similar Species: Females of *C. promethea* are similar in pattern, but their general coloring is more reddish to darker brown.

Description of Larva: Length to 70 mm (2 3/4 in.). The body is whitish green with a white low lateral stripe along the side. The second and third thoracic segments both have a pair of red dorsal scoli. The eighth abdominal segment has a single yellow dorsal scolus. Most abdominal segments have a dorsal, a lateral, and a ventral small black scolus.

Description of Pupa: The pupa is always to be found in a silken cocoon. The pupa is typical of the Saturniinae—brown, without a distinct cremaster, and smooth-bodied. The double-walled cocoon is spun within the folds of a leaf, but not fastened to the leaf stem, so that the cocoon eventually falls to the ground with leaf drop. The silk is dark, and the cocoon more or less irregularly shaped.

Habitat: This species feeds only on tulip tree foliage (*Liriodendron tulipifera*) in nature, and thus is to be found in forested regions where the plant grows.

Natural History: Bivoltine in Illinois. First-generation adults emerge in May from pupae that have overwintered on the ground within their cocoons. Both sexes are nocturnal. Mating occurs in the evening, usually before midnight. Oviposition begins the next night. Eggs are laid in small groups on the leaves of the host, whereas the early instars are gregarious, the later instars are solitary. Adults of the second generation are found in late summer. Their progeny, and usually some of the first-generation progeny, overwinter as pupae.

Status: Common at times, but scarce at the northern limits of its range. This species occurs where there are good stands of tulip trees. In Illinois, it is found in the southern part of the state, north up the Wabash Valley to Crawford County.

Tulip Tree Silkmoth

Reared male Tulip Tree Silkmoth, Pope County.

Reared female Tulip Tree Silkmoth, Pope County.

Field Guide to Silkmoths of Illinois

Reared Tulip Tree Silkmoth larvae (early instar), Pope County.

Reared Tulip Tree Silkmoth larvae (2nd instar), Pope County.

Reared Tulip Tree Silkmoth larva (5th instar, side view showing lateral stripe), Pope County.

Reared Tulip Tree Silkmoth cocoon, Pope County.

Reared Tulip Tree Silkmoth larva (5th instar, dorsal view), Pope County.

Tulip Tree Silk Moth

Reared male Tulip Tree Silkmoth (disturbed position), Pope County.

Reared male Tulip Tree Silkmoth (underside), Pope County.

Field Guide to Silkmoths of Illinois

Reared female Tulip Tree Silkmoth (disturbed position), Pope County.

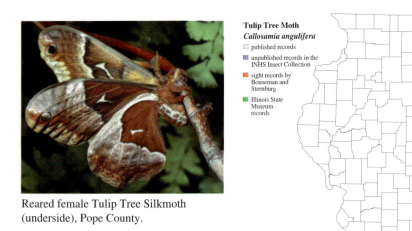

Reared female Tulip Tree Silkmoth (underside), Pope County.

Tulip Tree Moth
Callosamia angulifera

▧ published records

■ unpublished records in the INHS Insect Collection

■ sight records by Bouseman and Sternburg

■ Illinois State Museum records

Cecropia Moth or Robin Moth
Hyalophora cecropia (Linnaeus, 1758)

Subfamily: Saturniinae

This imposing moth is probably recognizable and known to more people than any other North American moth species. It is common over much of its range, and because of the relatively easy collection of its conspicuous winter cocoons, it is often brought into the classroom for study. In size, its wings have a greater total expanse in area than any of our other species, although the Cecropia may be exceeded in wingspan as measured from wing tip to wing tip by the Regal Moth.

Description of Adult: Wingspan 110–150 mm (4 1/4–6 in.) occasionally larger. One female specimen in the INHS collection has a total extended wingspan of 7 1/8 in. Sexes are alike in color. As is typical of the family, males have more broadly quadripectinate antennae than the females. The apex of the forewing has a violet patch. Behind this is a black spot with a blue inner margin. The outer margins of the wings are tan or grayish buff. A postmedial broad white line runs across both the fore and hindwing. Basad of the postmedial line the wings are black, dusted with a scattering of white scales imparting a grizzled appearance. Each wing has a lunate eyespot, often white, but sometimes suffused with orange-red. Distad of the postmedial line each wing has a broad red band, and beyond this a black band. Some individuals have the basal half of each wing broadly shaded with red-brown scales. All in all, this is a striking and magnificent insect. In addition to the wing colors, the body is red with white abdominal circular bands.

Similar Species: The Columbia Moth is much smaller and lacks red markings.

Description of Larva: Length to 100 mm (4 in.). First larval instar is black. The second instar is yellow. The third, fourth, and fifth instars are green with red or orange paired scoli on the meso- and metathoracic segments. All dorsal abdominal scoli are yellow. Lateral scoli are blue.

Similar Species: The Columbia Silkmoth has three pairs of red dorsal scoli instead of two pairs (two pairs thoracic, one pair first abdominal). Its scoli are more intensely red. Lateral scoli are white, not blue.

Description of Pupa and Cocoon: Pupae have a smooth brown integument. There is no cremaster. Cocoons are double-walled with an escape valve. Freshly spun cocoons are brownish, gradually becoming gray with time. Two forms occur: Most cocoons are elongate and quite firm, but some are baggy for reasons not understood; these have the outer layer loose, connected by silken threads passing through an open region between the outer and inner layers. The different types do not reflect sexual differences. Baggy cocoons are more frequent in brushy situations.

Habitat: Forests in rural and urban areas, most frequently in areas in the earlier stages of plant succession.

Natural History: Univoltine with a bimodal emergence. In east-central Illinois from 5 to 20 percent of overwintering pupae terminate diapause and begin to develop as soon as spring temperatures rise. These become adults in May. The remaining 80 to 95 percent require an additional warm period of a month or more before diapause ends and adult development occurs. The latter moths, which are most numerous, emerge in late June and early July. Laboratory experiments have shown that termination of diapause in the late emerging moths can be triggered by injection of the insect hormone ecdysone, thus bypassing the need for additional warm temperature to break diapause. Bimodal emergence is found in all Illinois populations of Cecropia Moth, although the time between emergences may vary. Under normal conditions, the progeny of one mated pair will be mixed, some early and some late. By selection during rearing, it is possible to develop strains with almost all individuals emerging early.

Emergence from the cocoon is usually in late morning. The moth remains quiet until dusk when the males begin a dispersal flight lasting an hour or more. The females remain still until pheromone release begins in the last few hours before dawn. It is during this time that males again become active, seeking females by flying upwind along odor trails to locate the signaling female. Mating takes place with no courtship. A mated pair, unless disturbed, will remain coupled until that evening, when they separate and oviposition begins. Eggs are laid in masses of 2 to about 10, normally on the larval host. However, in captivity a female will lay her eggs in an enclosure such as a paper bag, without any plant material present. Adults survive for a week to 12 days and then die as their stored energy reserves are depleted. First and second larval instars are gregarious but later instars disperse to feed solitarily. Larvae are very polyphagous. A partial list of the more important plants fed upon includes apple (*Malus* spp.), birch (*Betula* spp.), box elder (*Acer negundo*), buckthorn (*Rhamnus* spp.), dogwoods (*Cornus* spp.), garden peony (*Paeonia officinalis*), larch (*Larix* spp.), poplars (*Populus* spp.), rose (*Rosa* spp.), silver maple (*Acer saccharinum*), wild cherries (*Prunus* spp.), willows (*Salix* spp.), and many others. When fully grown, larvae empty the gut and actively search for a place to spin a cocoon. This will often be on a nearby plant, or in vegetation at the base of the larval host. Some larvae remain in the tree and spin there, but most cocoons are found hidden in tufts of grass or other plants at the base of the tree. The location of a cocoon is an important factor in survival. Cocoons visible on the branches or trunk of a woody plant are subject to winter predation by woodpeckers. Observation in Urbana-Champaign showed that 90 percent were attacked by woodpeckers during the winter and early spring months. Cocoons spun low and hidden were subject to predation in rural areas by white-footed mice. In the residential areas where house mice are dominant, low cocoons

Cecropia Moth

usually escaped predation, due to the inability of house mice to open cocoons. Cecropia Moth populations are often most dense in areas in an early successional stage, where trees and bushes are small. As trees mature, and with more birds present, Cecropia numbers decrease. The flight of the Cecropia is swift and powerful. We have records of flights of eight miles in one night by marked individuals from the release point. The species is thus well adapted to disperse and establish new populations in recently disturbed areas, in an early stage of succession. In Illinois, new residential areas, built on former cropland, often furnish ideal conditions for the Cecropia Moth.

Status: Statewide. Considered common, although relative to most insects their numbers are low. Their large size and bright colors often attract attention, and the ease with which cocoons can be found ensures familiarity with the species by many people.

Cecropia Moth eggs.

Reared Cecropia Moth larva (2nd instar), Champaign County.

Reared Cecropia Moth larva (4th instar), Champaign County.

Reared Cecropia Moth larva (5th instar), Champaign County.

Reared Cecropia Moth cocoon, Champaign County.

Reared Cecropia Moth cocoon slit open to show double wall, Champaign County.

Reared Cecropia Moth pupa, Champaign County.

Reared male Cecropia Moth, Champaign County.

Reared female Cecropia Moth, Champaign County.

Cecropia Moth

Reared male Cecropia Moth (disturbed position), Champaign County.

Reared female Cecropia Moth (disturbed position), Champaign County.

Reared Cecropia Moth cocoons (baggy form on left and compact form), Champaign County.

Field Guide to Silkmoths of Illinois

Reared female Cecropia Moth (resting position), Champaign County.

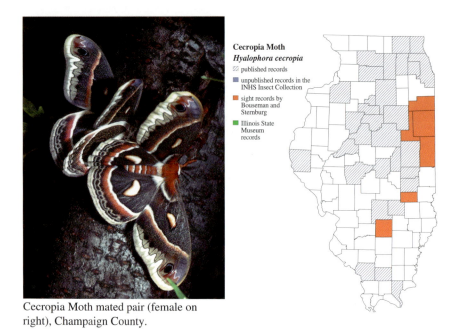

Cecropia Moth mated pair (female on right), Champaign County.

Cecropia Moth
Hyalophora cecropia
- published records
- unpublished records in the INHS Insect Collection
- sight records by Bouseman and Sternburg
- Illinois State Museum records

Columbia Silkmoth
Hyalophora columbia columbia (S.I. Smith, 1865)

Subfamily: Saturniinae

Note: Not recorded from Illinois. It occurs in Wisconsin and Michigan in tamarack bogs. These bogs occur in a few sites in northern Illinois.

Description of Adult: Wingspan 80–100 mm (3 1/8–4 in.). Sexes similar in color. The pattern is similar to that of the Cecropia Moth. A white postmedial line crosses both pairs of wings. The basal region is melanic or dark with a maroon color. The lunate discal eyespots are white. The wings distad of the postmedial line are blackish, with no trace of red.

Similar Species: Except for dwarfed individuals *H. cecropia* is much larger, and has prominent red postmedian bands, lacking in *H. c. columbia*.

Description of Larva: Length to 75 mm (3 in.). First instars are black, second instars are yellow, and later instars are green with paired red dorsal scoli on the mesothorax, the metathorax, and the first abdominal segments. The following dorsal scoli are yellowish. Lateral scoli are white with a black base.

Similar Species: Larvae of *H. cecropia* are larger when full grown, with only the two pairs of thoracic scoli reddish or orange. The other dorsal scoli are yellow, and the lateral scoli blue.

Description of Pupa and Cocoon: Pupation is within a double-walled silken cocoon with an escape valve. Its color is dark gray with silver and gold striations. It is often on the larval host, tamarack, but because the prepupal larva often wanders before spinning, it is often spun upon a nonhost plant. Cocoons are spun on the branches or the trunk if on a small tree. The pupa is smooth-bodied, with no cremaster, and is typical of the Saturniinae.

Habitat: Tamarack bogs (*Larix laricina*).

Natural History: The Columbia Silkmoth is closely associated with tamarack, the American larch, which is the primary larval host throughout the moth's range. There are reports from a limited number of sites of feeding by larvae on pin cherry (*Prunus pennsylvanicus*), alder (*Alnus rugosa*), and white birch (*Betula papyrifera*). In captivity, larvae are said to accept wild black cherry (*Prunus serotina*), and they do very well on European larch (*Larix decidua*). JGS and colleages maintained a captive culture on the latter for four generations in the 1970s. The species is univoltine throughout its range from Maine and Quebec west to the Plains and south to northern Wisconsin, Michigan, New York, and east. Emergence in the spring is early, usually in May or

June depending upon latitude. The flight season lasts several weeks, the timing varying within any site depending on local weather patterns. Emergence from the cocoon is in midmorning. Males search for females near dawn, at which time mating takes place. The mated pair then remains quiet until that night, when they separate. The females then begin oviposition; the eggs are laid singly or in small groups on the bases of the larch needles. They hatch in 10 days or so. The larvae disperse to feed singly until mature in mid to late summer. There are five instars. As stated above, cocoons are spun on the branches of the larval host, or on nearby vegetation.

Status: The Columbia Silkmoth is not known to occur in Illinois, but is included here for comparison and because it occurs in northern Wisconsin and in lower Michigan. There are a few tamarack bogs in northern Illinois and a chance stray female could conceivably produce a temporary population.

Remarks: The subspecies *H.c. gloveri* (Strecker, 1872) occurs west of the range of *H .c. columbia* and south through the Rocky Mountains. Its larvae feed on willows and poplars.

Reared Columbia Silkmoth larva (1st instar), Lower Michigan.

Reared Columbia Silkmoth larvae (2nd instar), Lower Michigan.

Reared Columbia Silkmoth larva (5th instar, side view), Lower Michigan.

Reared Columbia Silkmoth larva (5th instar, dorsal view), Lower Michigan.

Reared Columbia Silkmoth cocoon, Lower Michigan.

Columbia Silkmoth

Reared male Columbia Silkmoth, Lower Michigan.

Reared female Columbia Silkmoth, Lower Michigan.

Field Guide to Silkmoths of Illinois

Reared female Columbia Silkmoth (disturbed position), Lower Michigan.

Reared female Columbia Silkmoth (resting postion), Lower Michigan.

Columbia Silkmoth

Columbia Silkmoth mate pair (female above), Lower Michigan.

Ailanthus Silkmoth
Samia cynthia (Drury, 1773)
Subfamily: Saturniinae

Description of Adult: Sexes alike. Wingspan 100–115 mm (4–4 1/2 in.). Wings olive-brown with a lunate whitish discal spot on each wing. A sinuous white postmedial line, bordered on the basal side with black and outwardly with pink, runs across the fore and hindwings. There is a violet area at the apex of the forewing and a black-and-white eyespot. The abdomen has rows of white spots. The wings of the male are more pointed than the broader, more rounded wings of the female. The rami of the quadripectinate antennae are longest in the male.

Similar Species: None in Illinois.

Description of Larva: Length to 75 mm (3 in.). Color of larger larvae bluish green. Body with black dots and black spiracles, with three rows of scoli, dorsal, supraspiracular and subspiracular. On the later instars a whitish bloom covers the scoli and sometimes extends over the body.

Description of Pupa and Cocoon: The pupa is typical of the attacine saturniids. It is protected by a double-walled silken cocoon spun among the leaflets of an *Ailanthus* leaf. When the leaf stem eventually falls the cocoon drops to the ground, where it overwinters.

Habitat: Urban areas where tree-of-heaven (*Ailanthus altissima*) grows, typically railroad yards, against city fences, in vacant lots, depressed areas, city parks.

Natural History: Either univoltine or bivoltine depending on geographic location. The species overwinters as a pupa within a cocoon on the ground. Adults emerge in May and June. Eclosion of the adults is during the late morning hours. Mating takes place that day after dark. Oviposition begins the next evening. Larvae feed on the leaves of tree-of-heaven (*Ailanthis altissima*), rarely on other species.

Status: Not known to be established in Illinois. However, the species has been reported from Kentucky and St. Louis, Missouri. We know of no authentic Illinois specimens, but include the species because of the possibility of introduction. Occasional releases by breeders may occur, and are said to have occurred historically.

Remarks: The Ailanthus Silkmoth was deliberately released around 1860 in Philadelphia and other cities for the purpose of sericulture, an industry that failed in the U.S. The species is reared for its silk to a limited extent in Asia. Its present distribution in the U.S. appears to be declining.

Ailanthus Silkmoth

Reared male Ailanthus Silkmoth, from commercial source.

Female Ailanthus Silkmoth, Brooklyn, NY.

Field Guide to Silkmoths of Illinois

Male Ailanthus Silkmoth.

Reared Ailanthus Silkmoth larva (late instar), from commercial source.

Glossary

Allopatric	not occurring together geographically
Amplexiform	wing coupling for flight by means of overlapping wings
Attacine	belonging to the saturniid tribe Attacini
Basad	toward the base of an appendage or toward the body
Batesian mimicry	mimicry when the model is unpalatable, poisonous, or dangerous, and the mimic is palatable and harmless
Bipectinate	with each antennal segment bearing two branches
Bivoltine	two generations per year
Cremaster	a hooked structure at the tip of the last segment of a pupa
Crochets	small, hard cuticular hooks on the larval prolegs used to cling to surfaces
Diapause	arrested development, in response to environmental cues and controlled by hormonal changes
Diurnal	active during daylight hours
Dimorphic	with two forms
Discal cell	a cell at the base of a wing
Ecdysis	process of shedding the insect's cuticle
Ecdysone	a hormone controlling metamorphosis
Eclosion	emergence, as from a shed insect cuticle or cocoon
Frenulum	a bristle or bristles of the hindwing used to couple the wings in flight
Humeral angle	expanded basal margin of hindwing often involved in amplexiform wing coupling

Instar	the insect itself between molts
Labium	fused second maxillae, the appendage of the fourth postoral head segment
Mandible	one of a pair of appendages of the second, postoral head segment used in feeding primitively for chewing
Maxilla	one of a pair of appendages of the third postoral head segment, used as an adjunct to the mandibles in feeding
Medial	located midway from the wing base
Müllerian mimicry	when two or more unpalatable, poisonous, or dangerous species have evolved similar form, color, or behavior
Multivoltine	two or more generations per year
Nocturnal	active at night
Pheromone	a substance produced by one individual that causes a response by another individual within the same species
Polyphagous	feeding on many different plant species
Postmedial	located distad of the medial area of a wing
Proleg	an abdominal leg of a lepidopterous larva, bearing distal hooks
Quadripectinate	with four branches per antennal segment
Ramus	one of the branches of a pectinate antennal segment
Retinaculum	a structure of the forewing that functions as a clasp for the frenulum during wing coupling
Scolus	a tubercle bearing setae or spines found on lepidopterous larvae
Stemmata	(singular stemma) larval eyes of a caterpillar, usually in a semicircle of six on the lower sides of the cranium
Sympatric	occurring together geographically

Species Checklist

For common names we have followed Covell, *A Field Guide to the Moths of Eastern North America*. We have used the generic name *Syssphinx*, following Lemaire.

Family Saturniidae

    Subfamily Ceratocampinae

        _____ *Citheronia regalis* (Fabricius, 1793)
                Regal Moth or Royal Walnut Moth

        _____ *Citheronia sepulchralis* Grote and Robertson, 1865
                Pine-devil Moth

        _____ *Eacles imperialis* (Drury, 1773)
                Imperial Moth

        _____ *Anisota stigma* (Fabricius, 1775)
                Spiny Oakworm Moth

        _____ *Anisota virginiensis* (Drury, 1773)
                Pink-striped Oakworm Moth

        _____ *Anisota senatoria* (J.E. Smith, 1797)
                Orange-striped Oakworm Moth

        _____ *Dryocampa rubicunda* (Fabricius, 1793)
                Rosy Maple Moth

        _____ *Syssphinx bisecta* (Lintner, 1879)
                Bisected Honey Locust Moth

        _____ *Syssphinx bicolor* (Harris, 1841)
                Honey Locust Moth

    Subfamily Hemileucinae

        _____ *Hemileuca maia maia* (Drury, 1773)
                Buck Moth

_____ *Hemileuca nevadensis* Stretch, 1872
Nevada Buck Moth

_____ *Automeris io* (Fabricius, 1775)
Io Moth

Subfamily Saturniinae

_____ *Antheraea polyphemus* (Cramer, 1776)
Polyphemus Moth

_____ *Actias luna* (Linnaeus, 1758)
Luna Moth

_____ *Callosamia promethea* (Drury, 1773)
Promethea Moth or Spicebush Silkmoth

_____ *Callosamia angulifera* (Walker, 1885)
Tulip Tree Silkmoth

_____ *Hyalophora cecropia* (Linnaeus, 1758)
Cecropia Moth or Robin Moth

_____ *Hyalophora columbia columbia* (S.I. Smith, 1865)
Columbia Silkmoth

_____ *Samia cynthia* (Drury, 1773)
Ailanthus Silkmoth

Additional Reading

Field Guides

Covell, C.V., Jr. 1984. A field guide to the moths of eastern North America. Houghton Mifflin Company, Boston. xv + 496 pp.

Kricher, J.C., and G. Morrison. 1980. A field guide to eastern forests. Houghton Mifflin Company, Boston. xviii + 368 pp.

Lutz, F.E. 1948. Field book of insects of the United States and Canada, aiming to answer common questions. 3d ed. G.P. Putnam's Sons, New York. 510 pp.

Milne, L., and M. Milne. 1980. National Audubon Society field guide to North American insects and spiders. Alfred A. Knopf, New York. 989 pp.

Faunal Works

Covell, C.V., Jr. 1999. The butterflies and moths (Lepidoptera) of Kentucky: an annotated checklist. Kentucky State Nature Preserves Commission Scientific and Technical Series No. 6. xiv + 220 pp.

d'Abrera, B. 1995. *Saturniidae mundi*. Saturniid moths of the world. Part I. Automeris Press, Keltern, Germany. 177 pp.

d'Abrera, B. 1998. *Saturniidae mundi*. Saturniid moths of the world. Part III. Goecke & Evers, Keltern, Germany. 171 pp.

Heitzman, J.R., and J.E. Heitzman. 1987. Butterflies and moths of Missouri. Missouri Department of Conservation, Jefferson City. viii + 385 pp.

Metzler, E.H. 1980. Annotated checklist and distribution maps of the royal moths and giant silkworm moths (Lepidoptera: Saturniidae) in Ohio. Ohio Biological Survey Biological Notes No. 14. iv + 10 pp.

Opler, P.A. 1995. Lepidoptera of North America. 1. Distribution of silkmoths (Saturniidae) and hawkmoths (Sphingidae) of eastern North America. Contributions of the C.P. Gillette Museum of Insect Biodiversity, Department of Entomology, Colorado State University, Fort Collins. (Unpaginated).

Classification

Ferguson, D.C. 1971. Fascicle 20.2A, Bombycoidea (in part). Pages 1–153 *in* R.B. Dominick et al. The moths of North America north of Mexico. E.W. Classey, London.

Ferguson, D.C. 1972. Fascicle 20.2B, Bombycoidea (in part). Pages 155–275 *in* R.B. Dominick et al. The moths of North America north of Mexico. E.W. Classey, London.

Lemaire, C. 1978. Les Attacidae américains. Attacinae. Lemaire, Neuilly-sur-Seine. 238 pp. + 49 pls.

Lemaire, C. 1988. Les Saturniidae américaines. Ceratocampinae. Museo Nacional de Costa Rica, San Jose. 480 pp. + 64 pls.

Lemaire, C., and J. Minet. 1999. The Bombycoidea and their relatives. Pages 321–353 *in* N.P. Kristensen, ed. Handbook of zoology, Vol. IV, Arthropoda: Insecta. Part 35, Lepidoptera, moths and butterflies. Vol. 1: Evolution, systematics, and biogeography. Walter de Gruyter, Berlin and New York.

Michener, C.D. 1952. The Saturniidae (Lepidoptera) of the Western Hemisphere. Morphology, phylogeny, and classification. Bulletin of the American Museum of Natural History 98:335–502.

Natural History

Ford, E.B. 1972. Moths. 3d ed. Collins, London. xix + 266 pp.

Klots, A.B. 1958. The world of butterflies and moths. McGraw-Hill Book Company. New York, Toronto, London. 207 pp.

Leverton, R. 2001. Enjoying moths. T. & A.D. Poyser Ltd., London. xi + 276 pp.

Scoble, M.J. 1992. The Lepidoptera: form, function, and diversity. Oxford University Press, New York and Oxford. xi + 404 pp.

Tuskes, P.M., J.P. Tuttle, and M.M. Collins. 1996. The wild silk moths of North America: a natural history of the Saturniidae of the United States and Canada. Cornell University Press, Ithaca and London. ix + 250 pp.

Vesco, J.P. 2001. Moths and butterflies. Viking Studio, Penguin Putnam, New York. 183 pp.

Young, M. 1997. The natural history of moths. T. & A.D. Poyser Ltd., London. xiv + 271 pp.

Immature Stages

Godfrey, G.L., M. Jeffords, and J.E. Appleby. 1987. Saturniidae (Bombycoidea). Pages 513–521 *in* F.W. Stehr, ed. Immature insects. Vol. I. Kendall Hunt Publishing Company, Dubuque, Iowa. xiv + 754 pp.

Mosher, E. 1969. Lepidoptera pupae. Five collected works on the pupae of North American Lepidoptera. Entomological Reprint Specialists, East Lansing, Michigan. vii + 323 pp.

Tietz, H.W. 1972. An index to the described life histories, early stages and hosts of the Macrolepidoptera of the continental United States and Canada. A.C. Allyn, Sarasota, Florida. iv + 1041 pp. (Bound in two volumes).

Wagner, D.L., V. Giles, R.C. Reardon, and M.L. McManus. 1997. Caterpillars of eastern forests. United States Department of Agriculture Forest Service, Forest Health Technology Enterprise Team Publ. No. 96-34. United States Government Printing Office, Washington, DC. 113 pp.

Rearing

Collins, M.M., and R.D. Weast. 1961. Wild silk moths of the United States. Saturniinae. Collins Radio Company, Cedar Rapids, Iowa. iii + 138 pp.

Stone, S.E. 1991. Foodplants of world Saturniidae. The Lepidopterists' Society Memoir 4. xv + 186 pp.

Villiard, P. 1969. Moths and how to rear them. Funk & Wagnalls, New York. xiii + 242 pp.

Winter, W.D., Jr. 2000. Basic techniques for observing and studying moths and butterflies. The Lepidopterists' Society Memoir 5. xviii + 444 pp.

Economic Considerations

Johnson, W.T., and H.H. Lyon. 1991. Insects that feed on trees and shrubs: an illustrated practical guide. 2d ed., revised. Comstock Publishing Associates, Ithaca, New York. 560 pp.

Metcalf, R.L., and R.A. Metcalf. 1993. Destructive and useful insects: their habits and control. 5th ed. McGraw-Hill. New York. xv + unpaginated.

United States Department of Agriculture Forest Service. 1985. Insects of eastern forests. USDA Forest Service Miscellaneous Publication No. 1426. United States Government Printing Office, Washington, DC. x + 608 pp.

Photography

Angel, H. 1982. The book of nature photography. Alfred A. Knopf, New York. 168 pp.

Fitzharris, T. 1990. The Audubon Society guide to nature photography. Little, Brown and Company, Boston, Toronto, London. 167 pp.

Shaw, J. 1984. The nature photographer's complete guide to professional field techniques. Watson-Guptill Publications, New York. 144 pp.

Shaw, J. 1987. John Shaw's close-ups in nature. Watson-Guptill Publications, New York. 144 pp.

Shaw, J. 1991. John Shaw's focus on nature. Watson-Guptill Publications, New York. 144 pp.

Botanical References

Braun, E.L. 1967. Deciduous forests of eastern North America. Hafner Publishing Company, New York and London. xiv + 596 pp.

Leopold, D.J., W.C. McComb, and R.N. Muller. 1998. Trees of the central hardwood forests of North America: an identification and cultivation guide. Timber Press, Portland, Oregon. 469 pp.

Mohlenbrock, R.H. 1986. Guide to the vascular flora of Illinois. Revised and enlarged ed. Southern Illinois Press, Carbondale and Edwardsville. viii + 507 pp.

Mohlenbrock, R.H. 1996. Forest trees of Illinois. 8th ed. Illinois Department of Natural Resources, Springfield. 331 pp.

Petrides, G.A. 1988. A field guide to eastern trees. Houghton Mifflin Company, Boston. xv + 272 pp.

Recreational Reading

Cody, J. 1996. Wings of paradise: the great saturniid moths. University of North Carolina Press, Chapel Hill & London. xix + 163 pp.

Holland, W.J. 1903. The moth book: a popular guide to a knowledge of the moths of North America. Doubleday, Page & Company, New York. xxiv + 479 pp.

Matthews, P. 1957. The pursuit of moths and butterflies: an anthology. Chatto & Windus, London. 141 pp.

Stratton-Porter, G. 1912. Moths of the Limberlost. Doubleday, Page & Company, Garden City, New York. xiv + 370 pp.

Waldbauer, G. 1996. Insects through the seasons. Harvard University Press, Cambridge, Massachusetts and London, England. xiii + 289 pp.

De Omnibus Rebus

Boettner, G.H., J.S. Elkinton, and C.J. Boettner. 2000. Effects of a biological control introduction on three nontarget native species of saturniid moths. Conservation Biology 14(6):1798–1806.

Hodges, R.W., et al.,eds. 1983. Check list of the Lepidoptera of America north of Mexico. E.W. Classey Limited, London, and The Wedge Entomological Research Foundation, Washington, DC. xxiv + 284 pp.

Jeffords, M.R., G.P. Waldbauer, and J.G. Sternburg. 1980. Determination of the time of day at which diurnal moths painted to resemble butterflies are attacked by birds. Evolution 34(6):1205–1211.

Jensen, M.N. 2000. Silk moth deaths show perils of biocontrol. Science 290:2230–2231.

Sternburg, J.G., and G.P. Waldbauer. 1969. Bimodal emergence of adult cecropia moths under natural conditions. Annals of the Entomological Society of America 626(6):1422–1429.

Sternburg, J.G., and G.P. Waldbauer. 1978. Phenological adaptations in diapause termination by cecropia from different latitudes. Entomologia Experimentalis et Applicata 23:48–54.

Sternburg, J.G., and G.P. Waldbauer. 1984. Diapause and emergence patterns in univoltine and bivoltine populations of promethea (Lepidoptera: Saturniidae). The Great Lakes Entomologist 17(3):155–161.

Sternburg, J.G., G.P. Waldbauer, and M.R. Jeffords. 1977. Batesian mimicry: selective advantage of color pattern. Science 195:681–683.

Sternburg, J.G., G.P. Waldbauer, and A.G. Scarbrough. 1981. Distribution of cecropia moth (Saturniidae) in central Illinois: a study in urban ecology. Journal of the Lepidopterists' Society 35(4):304–320.

Waldbauer, G.P., and J.G. Sternburg. 1967. Differential predation on cocoons of *Hyalophora cecropia* (Lepidoptera: Saturniidae) spun on shrubs and trees. Ecology 48(2):312–315.

Waldbauer, G.P., and J.G. Sternburg. 1973. Polymorphic termination of diapause by cecropia: genetic and geographical aspects. Biological Bulletin 145:627–641.

Waldbauer, G.P., and J.G. Sternburg. 1978. The bimodal termination of diapause in the laboratory by *Hyalophora cecropia*. Entomologia Experimentalis et Applicata 23:121–130.

Index

Acer 24, 34, 46, 49, 59
Acer negundo 70
Acer rubrum 34
Acer saccharinum 34, 70
Acer saccharum 34
Actias luna 6, 53–57
Ailanthus altissima 80
Ailanthus Silkmoth 80–82
Alder 75
Alnus rugosa 75
Anisota 6, 28
Anisota senatoria 30, 32–33
Anisota stigma 28–30, 32
Anisota virginiensis 30–32
Antennae 3, 6–7
Antheraea polyphemus 6, 48–52
Apatelodinae 3
Apple 45, 49, 70
Ash 59
Automeris io 45–47
Basswood 45
Batesian mimicry 10, 58–59, 83
Battus philenor 10, 58–59
Betula 24, 43, 49, 70
Betula papyrifera 54, 75
Betula pumila 43
Birches 24, 49, 70
Bisected Honey Locust Moth 36–38
Blackberry 45
Black cherries 45, 59, 75
Bogbean 43
Bombycidae 3
Bombycoidea 3
Box elder 70
Buck Moths 2–4, 6–7, 41–43
Buckthorn 70
Callosamia angulifera 58, 64, 68
Callosamia promethea 10, 58, 63, 64
Carya 18, 46, 53
Cecropia Moth 7, 8, 11–12, 69–75
Celtis 45

Ceratocampinae 6, 17, 22, 24, 28, 30, 32, 34, 36, 38
Cercis canadensis 46
Cherries 49, 70
Citheronia regalis 17–18, 21–22
Citheronia sepulchralis 22
Classification 3–4
Columbia Silkmoth 5, 69, 75–79
Corn 45
Cornus 49, 70
Corylus 28, 41
Diospyros virginiana 18, 54
Distribution of Silkmoths 2
Dogwood 49, 70
Dryocampa rubicunda 10, 34–35
Dwarf birch 43
Eacles 6
Eacles imperialis 6, 24–27
Economic Considerations 2
Elm 45, 49
European larch 75
Family Saturniidae 1–3
Fraxinus 59
Garden peony 70
Geometrid 6
Geometridae 6
Gleditsia triacanthos 36, 38
Green-striped Mapleworm 35
Gymnocladus dioicus 36, 38
Hackberry 45
Hawkmoths 2
Hazel 28, 41
Hemileuca 3, 41, 43
Hemileuca maia maia 10, 41–43
Hemileuca nevadensis 10, 41–44
Hemileucinae 6, 41, 43, 45
Hickory 18, 46, 53
Hickory Horned Devils 17
Honey Locust Moth 38–40
Honey locust tree 36, 38
Hyalophora cecropia 69–74
Hyalophora columbia columbia 75–79
Hyalophora columbia gloveri 76
Hyparpax aurora 10, 34
Illinois Natural History Survey 2

Illinois State Museum 2
Imperial Moth 6–7, 24–27
Io Moth 2, 6–7, 45–47
Juglans 18, 24, 49, 53
Kentucky coffee tree 36, 38
Larch 70, 75
Larix 70, 75
Larix decidua 75
Larix laricina 75
Larval Morphology 4
Lepidoptera 3–4
Ligustrum 46
Lilac 59
Limacodidae 6
Lindera 59
Liquidambar styracflua 18, 54, 59
Liriodendron tulipifera 59, 64
Luna Moth 8–9, 53–57
Lythrum salicaria 43
Malus 45, 49, 70
Mandibles 4–5
Maples 24, 34, 46, 49, 59
Maps of Distribution 2
Menyanthes trifoliata 43
Mice 70–71
Mimicry 10, 34
Moon Moth 54
Moth Pupae 7
Müllerian mimicry 43
Neotropical Royal Moth 10
Nevada Buck Moth 41, 43–44
Notodontidae 34
Nymphalidae 6
Oak 24, 28, 30, 32, 34, 41, 43–44, 46, 49
Orange-striped Oakworm Moth 32–33
Paeonia officinalis 70
Papilio glaucus 59
Persimmon 18, 54
Pin cherry 75
Photographing Insects 12
Pine-devil Moth 22–23
Pink Prominent Moth 10, 34
Pink-striped Oakworm Moth 28, 30–32
Pinus 24

Pinus rigida 22
Pinus strobus 22
Pinus virginiana 22
Pipevine Swallowtail 10, 58–59
Plums 49, 70
Polyphemus Moth 7, 11, 48–52
Poplars 42–43, 46, 49, 70
Populus 42–43, 46, 49, 70
Primrose Moth 6, 10–11, 34
Privet 46
Promethea Moth 4, 8, 10–11, 58–64
Prunus 49, 70
Prunus pennsylvanicus 75
Prunus serotina 45, 59, 75
Psilopygida apollkinairei 10
Purple loosestrife 43
Quercus 24, 28, 30, 32, 34, 41, 46, 49
Quercus ilicifolia 41, 43
Redbud 46
Regal Moth 17–22, 69
Rhamnus 70
Rhus 18, 54, 59
Robin Moth 69
Rosa 49, 70
Rose 49, 70
Rosy Maple Moth 2, 34–35
Royal Walnut Moth 17
Rubus 45
Salix 42–43, 46, 49, 70
Salt Marsh Caterpillar Moth 42
Samia cynthia 80
Sassafras 24, 46, 59
Sassafras albidum 24, 46, 59
Saturniid 1–6, 11–13, 17, 30, 32, 34, 54, 80
Saturniid Relatives 14
Saturniidae 1–3, 6, 59
Saturniinae 6–8, 48, 53, 58, 64, 69, 75, 80
Schinia florida 10, 34
Scrub oak 43
Silver maple 70
Species Checklist 85
Sphingicampa 36
Sphingidae 2–3

Spicebush 59
Spicebush Silkmoth 58
Spiny Oakworm Moth 28–29
Subspiracular scoli 5
Sumac 18, 54, 59
Swallowtail 10, 58–59
Sweet gum 18, 54, 59
Syringa 59
Syssphinx bicolor 30, 36, 38–40
Syssphinx bisecta 36–38
Tamarack bogs 75–76
Tiger Swallowtail 59
Tilia 45
Tree-of-heaven 80
Tulip tree 59, 64
Tulip Tree Silkmoth 64–68
Ulmus 45, 49
Walnut 18, 24, 49, 53
White Birch 54, 75
Willows 42–43, 46, 49, 70
Wing vein terms 4
Zea mays 45

Notes